Assessing Stream Channel

Stability at Bridges in

Physiographic Regions

FOREWORD

The objective of this study was to expand and improve a rapid channel stability assessment method developed previously by Johnson et al. to include additional factors, such as major physiographic units across the United States, a greater range of bank materials and complexities, critical bank heights, stream types and processes, sand bed streams, and in-channel bars or lack of bars.[1] Another goal of this study was to tailor Thorne's reconnaissance method for bridge inspection and stability assessment needs.[2] Stream-bridge intersections were observed across the United States to develop and test the stability assessment method. Site visits were conducted at 57 stream-bridge intersections in 14 physiographic regions and subregions. Data collected and included in the report include locations and global positioning system (GPS) coordinates of the bridges, the physiographic Province, land use, stream classification, bed and bar material, percent of sand in the bed material, controls in the banks or on the bed, bank vegetation, bank material, bank height, and any erosion-related characteristics. Variability in stream types and common characteristics within each of the physiographic regions also were described. Thirteen indicators were identified for the stability assessment method. For each indicator, a rating of poor, fair, good, or excellent was assigned. An overall rank was obtained by summing the 13 ratings. To address sensitivities of various stream types to the indicators and rankings, the appropriate ranges of rankings were determined for three categories of stream channels. Each of the 57 stream-bridge intersections also was described in terms of lateral and vertical stability. Finally, a simplified version of Thorne's stream reconnaissance field sheets is presented for collecting data for the stream stability assessment and to provide a record of conditions at each visit.[2]

Gary L. Henderson
Director, Office of Infrastructure
Research and Development

NOTICE

QUALITY ASSURANCE STATEMENT

1. Report No. FHWA-HRT-05-072	2. Government Accession No.	3. Recipient's Catalog No.
4. Title and Subtitle Assessing Stream Channel Stability at Bridges in Physiographic Regions		5. Report Date July 2006
		6. Performing Organization Code
7. Author (s) Peggy A. Johnson		8. Performing Organization Report No.
9. Performing Organization Name and Address Department of Civil and Environmental Engineering Pennsylvania State University University Park, PA 16802		10. Work Unit No. (TRAIS)
		11. Contract or Grant No. DTFH61–03–P–00353
12. Sponsoring Agency Name and Address Office of Infrastructure Research and Development Federal Highway Administration Turner-Fairbank Highway Research Center 6300 Georgetown Pike McLean, VA 22101		13. Type of Report and Period Covered Final Report, August 2003 to August 2004
		14. Sponsoring Agency Code

15. Supplementary Notes
Contracting Officer's Technical Representative—J. Sterling Jones, HRDI–07; cosponsored by Jorge Pagan, Office of Bridge Technology

16. Abstract

The objective of this study was to expand and improve a rapid channel stability assessment method developed previously by Johnson et al. to include additional factors, such as major physiographic units across the United States, a greater range of bank materials and complexities, critical bank heights, stream types and processes, sand bed streams, and in-channel bars or lack of bars.[1] Another goal of this study was to tailor Thorne's reconnaissance method for bridge inspection and stability assessment needs.[2] Stream-bridge intersections were observed across the United States to develop and test the stability assessment method. Site visits were conducted at 57 stream-bridge intersections in 14 physiographic regions and subregions. Data collected and included in the report include locations and global positioning system (GPS) coordinates of the bridges, the physiographic Province, land use, stream classification, bed and bar material, percent of sand in the bed material, controls in the banks or on the bed, bank vegetation, bank material, bank height, and any erosion-related characteristics. Variability in stream types and common characteristics within each of the physiographic regions also were described. Thirteen indicators were identified for the stability assessment method. For each indicator, a rating of poor, fair, good, or excellent was assigned. An overall rank was obtained by summing the 13 ratings. To address sensitivities of various stream types to the indicators and rankings, the appropriate ranges of rankings were determined for three categories of stream channels. Each of the 57 stream-bridge intersections also was described in terms of lateral and vertical stability. Finally, a simplified version of Thorne's stream reconnaissance field sheets is presented for collecting data for the stream stability assessment and to provide a record of conditions at each visit.[2]

17. Key Words Bridge scour, stream stability, inspection, bridge maintenance, hydraulics	18. Distribution Statement		
19. Security Classif. (of this report) Unclassified	20. Security Classif. (of this report) Unclassified	21. No. of Pages 157	22. Price N/A

Form DOT F 1700.7   (8-72)   **Reproduction of completed page authorized**

SI* (MODERN METRIC) CONVERSION FACTORS

APPROXIMATE CONVERSIONS TO SI UNITS

Symbol	When You Know	Multiply By	To Find	Symbol
LENGTH				
in	inches	25.4	millimeters	mm
ft	feet	0.305	meters	m
yd	yards	0.914	meters	m
mi	miles	1.61	kilometers	km
AREA				
in^2	square inches	645.2	square millimeters	mm^2
ft^2	square feet	0.093	square meters	m^2
yd^2	square yard	0.836	square meters	m^2
ac	acres	0.405	hectares	ha
mi^2	square miles	2.59	square kilometers	km^2
VOLUME				
fl oz	fluid ounces	29.57	milliliters	mL
gal	gallons	3.785	liters	L
ft^3	cubic feet	0.028	cubic meters	m^3
yd^3	cubic yards	0.765	cubic meters	m^3
NOTE: volumes greater than 1000 L shall be shown in m^3				
MASS				
oz	ounces	28.35	grams	g
lb	pounds	0.454	kilograms	kg
T	short tons (2000 lb)	0.907	megagrams (or "metric ton")	Mg (or "t")
TEMPERATURE (exact degrees)				
°F	Fahrenheit	5 (F-32)/9 or (F-32)/1.8	Celsius	°C
ILLUMINATION				
fc	foot-candles	10.76	lux	lx
fl	foot-Lamberts	3.426	candela/m^2	cd/m^2
FORCE and PRESSURE or STRESS				
lbf	poundforce	4.45	newtons	N
lbf/in^2	poundforce per square inch	6.89	kilopascals	kPa

APPROXIMATE CONVERSIONS FROM SI UNITS

Symbol	When You Know	Multiply By	To Find	Symbol
LENGTH				
mm	millimeters	0.039	inches	in
m	meters	3.28	feet	ft
m	meters	1.09	yards	yd
km	kilometers	0.621	miles	mi
AREA				
mm^2	square millimeters	0.0016	square inches	in^2
m^2	square meters	10.764	square feet	ft^2
m^2	square meters	1.195	square yards	yd^2
ha	hectares	2.47	acres	ac
km^2	square kilometers	0.386	square miles	mi^2
VOLUME				
mL	milliliters	0.034	fluid ounces	fl oz
L	liters	0.264	gallons	gal
m^3	cubic meters	35.314	cubic feet	ft^3
m^3	cubic meters	1.307	cubic yards	yd^3
MASS				
g	grams	0.035	ounces	oz
kg	kilograms	2.202	pounds	lb
Mg (or "t")	megagrams (or "metric ton")	1.103	short tons (2000 lb)	T
TEMPERATURE (exact degrees)				
°C	Celsius	1.8C+32	Fahrenheit	°F
ILLUMINATION				
lx	lux	0.0929	foot-candles	fc
cd/m^2	candela/m^2	0.2919	foot-Lamberts	fl
FORCE and PRESSURE or STRESS				
N	newtons	0.225	poundforce	lbf
kPa	kilopascals	0.145	poundforce per square inch	lbf/in^2

*SI is the symbol for the International System of Units. Appropriate rounding should be made to comply with Section 4 of ASTM E380.
(Revised March 2003)

TABLE OF CONTENTS

LIST OF FIGURES

LIST OF TABLES

1. INTRODUCTION

The goal of bridge inspections is to assess the safety of bridges on a regular basis so that any deficiencies will be identified and corrected. Given the large number of bridges over water in any State, bridge inspectors must inspect the superstructure, substructure, and waterway of each bridge in a short amount of time. A typical range of time for bridge inspections is 15 minutes to 2 hours, depending on the complexity and condition of the bridge. A more detailed inspection might ensue if a deficiency is detected. In the case of waterways and erosion, a hydraulic engineer might visit the bridge to assess the situation in greater detail. For either of these levels of inspection, and given the very limited right-of-way at most bridges, the inspector or engineer typically will not walk more than a few hundred feet upstream or downstream. Most inspectors do not leave the bridge right-of-way. Thus, a method is needed for systematically assessing the stability of the stream channel with respect to the bridge. The ability to assess channel stability in the vicinity of bridges also is needed for designing road crossings, and for mitigating and predicting erosion at those structures. Bridge failures due to geomorphic or regional instability have been experienced in many locations in the United States and elsewhere. Federal Highway Administration (FHWA) guidelines for stream stability and erosion at bridges, such as Hydraulic Engineering Circular (HEC)-20[3] and HEC-18,[4] describe examples of problems at bridges caused by regional channel degradation and lateral bank changes. These guidelines require that engineers assess channel instability in their bridge assessments. However, for most bridges, only a preliminary assessment can be conducted due to time and money constraints.

The National Highway Institute (NHI) training course for bridge inspectors and hydraulic engineers has been based on a data collection method developed by Thorne.[2] The user completes a number of data sheets by collecting primarily qualitative geomorphic data. Although the method is very complete and provides a systematic method of collecting data at every site, there are several problems in its use in bridge inspections. First, there generally is not enough time to collect such detailed data, nor are most inspectors or even hydraulic engineers adequately trained to identify all of the factors. In addition, the level of data may not be necessary for the task at hand. Finally, after the data are collected, there is no systematic method for synthesizing the data for use in determining stream stability and decisionmaking.

Johnson et al. developed a rapid channel stability assessment method based on geomorphic and hydraulic indicators for use at bridges.[1] This method has been included in the most recent revision of HEC-20.[3] It is used in HEC-20 as a method to provide a semiquantitative level 1 analysis and to determine whether it is necessary to conduct a more detailed level 2 analysis. Thirteen qualitative and quantitative stability indicators are rated, weighted, and summed to produce a stability rating for gravel bed channels. The rapid stability method provides information that can aid in decisionmaking with respect to design, repair, rehabilitation, or replacement of a bridge or culvert. Given the Federal and State requirements of inspecting bridges for local, contraction, and regional scour, it is important to have a method in place that bridge engineers and inspectors can use to make initial judgments on regional channel instability that might be detrimental to a bridge.

The rapid assessment method developed by Johnson et al.[1] was based largely on previous assessment methods.[5,6,7]

1

Advantages of the method include:

- This method weights each criterion based on its impact on stream channel stability, giving lower weight to indicators, such as debris jam potential, and greater weight to indicators, such as mass wasting.
- The rapid assessment method does not have a single variable that can dominate the rating of channel stability.
- Evaluation of each indicator is categorized as excellent, good, fair, and poor with three values in each range.
- This method provides several quantitative indicators, such as bed shear stress ratio, while incorporating fewer ambiguous criteria, such as brightness and clinging aquatic vegetation, or criteria that are difficult to assess.
- The method includes bridge and culvert variables.

The assessment method was tested for selected streams in the Piedmont of Maryland and the Appalachian Plateau area of northern Pennsylvania. Since the assessment method was developed, a number of limitations have been identified, particularly when used outside of the area for which it was calibrated and tested.

One way to incorporate a large number of these complexities is to differentiate streams according to a chosen classification scheme. Montgomery and Buffington developed a stream classification scheme that is a function of processes that occur in various types of streams.[8,9] The Montgomery-Buffington stream classification scheme is based primarily on stream channel function rather than form. They categorize streams as braided, dune-ripple, pool-riffle, plane-bed, step-pool, cascade, bedrock, and colluvial. The indicators of stream type include typical bed material, bedform pattern, reach type (transport or type), dominant roughness elements, dominant sediment sources, sediment storage elements, typical slope, typical confinement, and pool spacing. They used this classification scheme to predict the response of a channel to changes in hydrology and sediment transport.

The U.S. Army Corps of Engineers (USACE) developed a classification scheme that is based essentially on the location and function of a stream within a watershed.[10] It is the only classification scheme that also includes altered streams. This method categorizes streams as mountain torrents, alluvial fans, braided rivers, arroyos, meandering alluvial rivers, modified, regulated, deltas, underfit streams, and cohesive streams. There are no quantitative thresholds for these streams; rather, qualitative characteristics of each stream type are given.

Many other classification schemes exist, but some require relatively large amounts of data that are time-consuming to collect and that do not necessarily provide information useful to a stability analysis. Combining several classification schemes, such as the USACE and Montgomery-Buffington schemes, may provide a basis for the classification of stable channel characteristics for different stream types.

OBJECTIVE

The objective of this study was to expand and improve the Johnson et al. rapid stability assessment method to include additional factors, such as major physiographic units across the

United States, range of bank materials and complexities, critical bank heights, stream type and processes, sand bed streams, and in-channel bars or lack of bars.[1] The assessment method was to be based on a similar format as Johnson et al., with improvements to be generally applicable in all types of streams across the United States.[1] The stream stability assessment method was also to be self-contained so that no additional data collection forms or methods were necessary. However, the use of forms that provide a systematic method for observations is desirable. Thus, the data collection was to be based on the reconnaissance method developed by Thorne.[2]

However, given that Thorne's method is very detailed and requires numerous data beyond that needed for bridge inspections and assessing stability, another goal of this study was to tailor Thorne's reconnaissance method for bridge inspection and stability assessment needs. The result of the project is a method to help bridge inspectors assess the stability of stream channels quickly at bridges that satisfy the following criteria:

- The method is based on the idea that only the channel stability in the short term is needed since inspectors check each waterway every 2 years.
- The method is based only on stability in the immediate vicinity of bridge (admittedly, this could overlook changes that can occur rapidly, such as knickpoint migration).
- The method must be quick and sufficiently accurate without time-consuming measurements, surveys, or calculations.

2. BACKGROUND AND LITERATURE REVIEW

A healthy, stable stream is resilient to disturbances, such as the passing of storm events and changes induced by humans. Dimensions of the stable stream channel are sustainable over decades. There is variability in roughness, which is important to ecological diversity. The stable stream is characterized by healthy, upright, woody vegetation; low banks that are not susceptible to mass wasting (gravity failures); and a flood plain that is connected to the river. Thus, during moderate flow events, the flood plain is active. Figure 1 provides an example of a stable stream. On the other hand, an unstable stream is characterized by overheightened, oversteepened banks that are susceptible to mass wasting, evidence of geotechnical failure planes along the banks, lack of diverse, upright woody vegetation, and the flood plain is disconnected from the channel so that moderate to high flows remain within the channel banks. Thus, wetlands tend to drain, and the nutrient source to the stream is cutoff. Figure 2 provides an example of an unstable stream channel.

Thorne et al. categorize alluvial channel stability as unstable, stable-dynamic, or stable-moribund.[7] They defined an unstable channel as one where degradation, aggradation, width adjustment, or planform changes were actively occurring in time and space. However, the main requirement is that there is net morphological change over engineering time scales. A dynamically stable channel is defined by Thorne et al. as one in which the characteristic dimensions do not change over engineering time scales.[7] Thorne et al. also define a moribund channel as one in which the characteristic dimensions have been formed by a prior flow regime different from that which is presently observed, or more likely, due to channel widening and dredging in low energy rivers.[7] Moribund channels are unlikely to recover from past engineering activity even if allowed to do so, because the river is unable to mobilize its bed material.

Brookes inferred channel stability in terms of stream power.[11] Based on field observations of stable and unstable streams in the United Kingdom, he found that in unconfined lowland, meandering channels—streams in which the stream power at bankfull discharge was greater than about 35 watts per square meter (W/m^2)—were unstable in terms of erosive adjustment. In these channels where stream power was less than 25 W/m^2, the stream was stable. Although such guidance is certainly useful, it is often very difficult to define bankfull in an unstable channel.[12] In addition, the criterion developed by Brookes will only be valid in the region where he collected the observations.

Figure 1. Stable stream in central Pennsylvania.

Figure 2. Unstable stream in western Pennsylvania.

Chorley and Kennedy described stability in terms of three types of equilibrium: (1) static, in which a static condition is created by a balance in opposing forces; (2) steady-state, in which the properties of a stream randomly oscillate about a constant state; and (3) dynamic, in which a balanced state is maintained by dynamic adjustments.[13] Richards showed that in a natural, stable channel, channel dimensions constantly adjust to passing floods.[14] So, although a stable channel has constant average dimensions over a medium timeframe (on the order of decades), those dimensions vary about the average value. Figure 3 shows an example of variation in width over time about the average width.

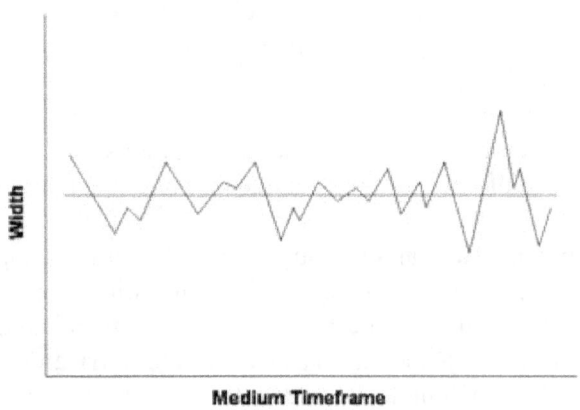

Figure 3. Variation of channel width over medium timeframe about
the stable mean (after reference 14).

Knox defines a stable stream as "one in which the relationship between process and form is stationary and the morphology of the system remains relatively constant over time."[15] At bridges, stability also implies limited lateral movement so that the channel is more or less centered beneath the bridge opening. A geomorphically stable channel that has considerable lateral migration is likely to be considered unstable by the engineer concerned with bridge safety.

Channel stability must be defined in terms of both time and space. The temporal and spatial scales used vary depending on the application. Temporal scales for channel stability can range from medium, in which one might be concerned about bridge safety or ecological recovery, to long term, which would include geomorphic and geologic stability. A short timeframe is considered to be on the order of 1 or 2 years; medium is decades to 100 years, typical of engineering design lives; and long term is hundreds to thousands of years. Spatial scales can also vary widely depending on how stability is defined. Length of stream over which stability is determined can be as short as several hundred feet, to 20 stream widths (a rule-of-thumb established by Leopold), to miles of stream.

CHANNEL ADJUSTMENTS

Rivers become temporarily unstable when new hydrologic or sediment load conditions are imposed.[16] Lane described this process as a proportionality between the loads entering the stream:[17]

$$Q_s d \, \% \, QS$$

(1)

where Q_s = sediment discharge, d = sediment size, Q = water discharge, and S = slope. Thus, a change in either of the loads, Q_s or Q, will result in adjustments of sediment size or slope. Hey expanded on equation 1 by determining the dependent variables that will adjust according to changes in the independent variables of the equation.[18] The independent variables are sediment discharge, bed and bank sediment characteristics, water discharge, and valley slope; the dependent variables include velocity, mean flow depth, channel slope, width, maximum flow depth, bedform wavelength, bedform amplitude, sinuosity, and meander arc length.

Changes in the independent variables can be brought about by either natural events or human-induced modifications. The changes can be direct or indirect. Natural events that increase sediment discharge include landslides and destabilization of channel banks by extreme hydrologic events. Water discharge is increased as storms and hurricanes create flooding in the stream channels and flood plains. Climatic changes also can gradually increase or decrease water discharge to a channel.

Human modifications to stream channels such as straightening, clearing, dredging, and widening can result in dramatic responses within the reach directly modified, as well as upstream or downstream of the modified reach. A good example of this is channel straightening. Straightening imposes an increased channel slope in the modified reach. To adjust to the new slope, a head cut often will proceed upstream, rapidly lowering the channel elevation. Many other modifications can affect the loads to a stream channel. Downstream of a dam, sediment discharge is decreased, typically resulting in bed degradation and a change in slope. Dams, other than run-of-the-river dams, also change the water discharge such that the discharge downstream is steadier at a higher discharge than for previous low flows. Larger events typically are stored in the reservoir. The result is downstream degradation due to maintaining a higher-than-normal flow over an extended period of time. Land use changes have significant, indirect impacts on channel adjustments. Deforestation for the purposes of either urbanization or agriculture often dramatically impact stream channels. Without woody vegetation, the banks become more susceptible to changes in discharge. Removing vegetation across the flood plain creates a reduced roughness and infiltration surface, thus increasing both the magnitude and timing of the flood hydrographs in the streams. This, in turn, increases movement of sediment in the river banks and bed. Construction during urbanization in a watershed increases fine sediment to a stream channel. Depending on the type of channel, this increase in sediment can change the channel morphology.

The response of a river to modifications in the sediment and water discharge depends on the type of channel and the type of modification. Changes in sediment or water discharges can occur as

8

either a pulse or a step (chronic) change.[19] A pulse may result in a temporary channel adjustment, but then return to its previous equilibrium dimensions. However, a step change is more likely to result in a permanent change to the stream stability and equilibrium dimensions. The length of time over which the channel reaches its new equilibrium or returns to a previous state depends on the intensity of the change in load as well as the type of channel and its resilience. Montgomery and MacDonald provide tables of the relative sensitivity of alluvial channel types to chronic changes in coarse sediment, fine sediment, and discharge.[20] The channel types are cascade, step-pool, plane bed, pool-riffle, and dune-ripple. For each channel type, they determine sensitivity to change as very responsive, secondary or small response, and little or no response. In every case, pool-riffle streams are the most sensitive to changes in the load. Their dimensions (depth and width) and bank stability are very sensitive to changes in coarse sediment supply and to increases in discharge. Bed material in these channels is also very responsive to changes in sediment supply and water discharge. By comparison, cascade and step-pool channels are not as sensitive and will maintain their dimensions and bank stability under conditions of change in sediment and water supply.

CHANNEL STABILITY AT BRIDGES

Knowledge of the spatial and temporal trends of channel adjustments is central to protecting and maintaining bridges. One well-known bridge collapse due to stream channel instability is the U.S. Route 51 bridge over the Hatchie River in Tennessee. During a 3-year flood, this bridge collapsed, killing eight people. The collapse was caused by lateral channel migration of 25.3 m over 13 years. The rate of lateral migration had increased dramatically following channel straightening to reduce the angle at which the channel approached the bridge. There are many other examples of bridge failures following channel modifications. Straightening of the Willow River in southwestern Iowa led to channel bed degradation and gully formation, resulting in the need to repair and reconstruct roads and bridges in the area.[21] Straightening and dredging of the Homochitto River in southwest Mississippi and the Blackwater River in Missouri caused significant bed degradation and widening, and led to the collapse of several bridges.[22,23] Additional bridge failures occurred in straightened western Tennessee channels as a result of channel bed degradation, channel widening, and local scour.[24,25]

Channel instability in the vicinity of a bridge can be arrested through the use of bank and bed stabilization structures, but if they fail during a hydrologic event, the bridge is at risk again. As an example, in 1995, a railroad bridge near Kingman, AZ, collapsed as an Amtrak® train crossed it, injuring more than 150 people. The cause was the sudden upstream migration of a head cut during heavy rains. Before this hydrologic event, the head cut migration had been halted by a check dam. When the check dam failed during the storm, the head cut was free to travel upstream.

Several studies have been conducted to assess the reliability of bridges in which piers and/or abutments are in an unstable, adjusting stream.[26,27] However, the key to assessing risk or reliability is identifying that a problem or potential for a problem exists and documenting the condition.

9

METHODS FOR COLLECTING STREAM CHANNEL DATA

Systematic data collection is an integral part of conducting a reconnaissance along a stream or assessing channel stability. The amount of data that is required depends on the level of detail desired. A wide range of data is useful in assessing stream channel conditions. The data include topographic maps, aerial photos, bridge inspection reports, hydrologic and hydraulic reports, stream gage data, and other geomorphic reports. Aerial photos and topographic maps are very useful in providing an overall view of the bridge, the stream below it, and the watershed conditions. Comparing photos and maps over a period of years is helpful in assessing rates of change, particularly at a larger scale. Both aerial photos and topographic maps can be viewed online at http://terraserver-usa.com. These tools help visualize the location of the bridge relative to the location of meanders, as well as the bridge alignment. Given the relative ease of checking aerial photos, this should be done as a standard part of any survey. In addition to studying photos and maps, examining previous reports on assessments conducted at or near the bridge is useful to determine trends. Given that bridge inspections are conducted at least every 2 years, typically with one or two cross sections measured, these are good reports to compare for changes over a longer period of time. Geomorphic assessments that have been conducted along the stream, although they may not be concerned with the bridge, are also excellent sources of information. HEC-20 details these types of data and where to access them.[3]

Collecting data along a stream to assess stream condition can include a wide variety of data and levels of detail. Thus, a systematic method of collection is essential to producing consistent data sets that can be compared and used for future analyses. The only complete, systematic, geomorphic data collection system that exists today is that created by Thorne.[2] In this system, multiple pages of forms provide a systematic methodology for collection of data and subjective observations. Data collection begins with geological and watershed level observations, then continues to focus on the stream corridor and hill slopes, and finally examines the actual bed and banks of the channel or water body. The data set developed through this reconnaissance provides complete documentation of current conditions. In addition, photographs are taken to help document current conditions. The Thorne reconnaissance method does not address infrastructure within a reach; thus, it is necessary to add parameters for that case. Johnson et al. revised the Thorne data sheets to suit streams in urban environments and provide descriptions of conditions at instream structures.[28] The data collected included descriptions of the valley, channel, bed sediment, bank material, vegetation, erosion, flood plain, instream structures, and reach measurements. According to Thorne, a reconnaissance could range from a very detailed study over 5–10 river widths that would include 1 pool-riffle couplet, individual meander, primary bifurcation-bar-confluence unit in braided channel, to a low level detail study over a much longer reach in which channel form and processes do not change significantly.[2]

The NHI training course on bridge scour and stream stability has emphasized the use of reconnaissance sheets developed by Thorne for systematically collecting geomorphic data.[3] The sheets are divided into five sections, with each section further divided into parts that focus on various aspects of the stream. These can be summarized as:

- Section 1—Scope and Purpose, including general information on the location and details of the project.

- Section 2—Region and Valley Description, including descriptions of the area around the river valley, river valley and valley sides, flood plain, vertical relation of channel to valley, and lateral relation of channel to valley.
- Section 3—Channel Description, including a description of the channel dimensions, controls, bed materials, and bar types and materials.
- Section 4—Left Bank Survey, including bank characteristics, vegetation, bank erosion, geotechnical failures, and toe sediment accumulation of the left bank.
- Section 5—Right Bank Survey, including bank characteristics, vegetation, bank erosion, geotechnical failures, and toe sediment accumulation of the right bank.

The method also includes numerous entries for subjective or interpretive observations. Bridge abutments and armor protection are entered on the data sheets as obstructions. Although individual items on the data sheets may not indicate channel stability or instability directly, the data are collectively important in assessing long-term stability. Table 1 provides the relationships between the data collected in Thorne's reconnaissance and long-term indications of stability. These relationships provide impetus for the development of the simplified reconnaissance sheets in terms of using the stability assessment method described in this report.

Table 1. Data items from reconnaissance sheets related to stream stability indicators.

Section and Part Number	Reconnaissance Parameter	Relationship to Stability
Section 1	General information	Basic information on project; should include bridge number
Section 2, Part 1. Area around river valley	Terrain	Important in some classification methods
	Drainage pattern	Descriptive of setting
	Surface geology	Sediment source information, but difficult to identify
	Rock type	Affects erosion rate to some extent, generally difficult to identify during short visit
	Land use	Important to hydrologic response and erosion rates
	Vegetation	Important to hydrologic response and erosion rates
Section 2, Part 2. River valley and sides	Location of river	Indicative of large scale channel behavior
	Valley shape	Large scale descriptor
	Valley height	Along with angle, may be important for indicating sediment source, such as land sliding into river
	Side slope angle	Along with height, may be important for indicating sediment source, such as land sliding into river
	Valley side failures	Sediment source
	Failure locations	Indicates whether potential sediment loadings are upstream or downstream of the bridge
Section 2, Part 3. Flood plain	Valley floor type and width	Indicative of confinement and lateral stability
	Surface geology	Indicates erosion rates
	Land use	Ongoing changes in land use critical to stability
	Vegetation	Important to whether sediment sources are protected
	Buffer strip and width	Important to lateral stability and erosion rates
	Left and right overbank	Roughness used to assess hydraulics of flow

Table 1. Data items from reconnaissance sheets related to stream stability indicators, continued.

Section and Part Number	Reconnaissance Parameter	Relationship to Stability
Section 2, Part 4. Vertical relation of channel to valley	Terraces	Identifies previous incision
	Trash lines	Identifies high water surface elevations
	Overbank deposits	Provides sense of median size of sediment discharge
	Levees	Increases shear stress along bottom; prohibits flood plain activity
Section 2, Part 5. Lateral relation of channel to valley	Planform	Type and dimensions of meanders indicate relative rate of lateral moving
	Flood plain features	Provide evidence of prior meander movement
Section 3, Part 6. Channel description	Dimensions	Width and depth can indicate entrenchment and stability; slope indicates stream power
	Flow type	Indicates flow energy
	Bed controls and types	Effects vertical stability or where erosion might occur
	Width controls and types	Effects lateral stability or where erosion might occur
Section 3, Part 7. Bed sediment description	Bed material	Overall size category indicates energy of stream to transport
	Bed armor	A type of vertical control
	$D_{50,84,16}$	Use to compute critical shear stress
	Substrate size	Indicates material available to movement after armor removed
	Sediment depth	Depth of mobile sediment
	Bedforms	Can significantly effect flow resistance
	Bar types and sediment size	Number, size, location, vegetation, and overall sediment size indicative of vertical changes

Table 1. Data items from reconnaissance sheets related to stream stability indicators, continued.

Section and Part Number	Reconnaissance Parameter	Relationship to Stability
Section 4 and 5, Parts 8 and 13. Left bank characteristics	Type	Layering and cohesiveness key to bank stability
	Bank materials	Level of cohesiveness controls stability
	Protection	Important to stabilization of bank materials
	Layer thickness	Along with layer materials important to stability
	Bank height	Along with angle, height indicates mass wasting potential
	Bank slope	Along with height, angle indicates mass wasting potential
	Profile shape	Indicates potential for geotechnical failures
	Tension cracks	Indicates potential for geotechnical failures
Sections 4 and 5, Parts 9 and 14. Left bank face vegetation	Vegetation	Plays key role in stabilizing banks and slowing lateral erosion
	Orientation	Indicates rate of bank movement
	Tree types	Roots of different trees better for holding soil in place and providing drainage
	Density and spacing	Important to how much erosion control exerted
	Location, health, diversity, height	Indicates growing rates, and therefore, erosion rates
Sections 4 and 5, Parts 10 and 15. Bank erosion	Fluvial erosion location	Indicates whether erosion activity is occurring in specific locations indicating problems
	Erosion status and rate	Difficult to assess on short site visit, but indicates stabilizing versus destabilizing conditions
Sections 4 and 5, Parts 11 and 16. Bank geotechnical failures	Failure location	Indicates areas that are contributing to destabilization
	Present status	Difficult to assess on short site visit, but indicates stabilizing versus destabilizing conditions
	Failure scars	Indicates prior mass wasting
Sections 4 and 5, Parts 12 and 17. Left bank toe sediment accumulation	Stored bank debris	Indicative of bank material and failure mechanisms
	Vegetation	Indicative of stabilizing toe
	Age, health, and type of vegetation	Important to showing stability

CHANNEL STABILITY ASSESSMENT METHODS

A number of methods currently are available for assessing channel stability. Some require the expertise of an experienced geomorphologist, while others require only a brief period of training. All of these methods are, at least in part, based on observations of a variety of parameters that describe the characteristics and conditions of the channel and surrounding flood plain. The purpose of each of these methods is to assess the current condition of the channel and possibly identify the processes that are acting to change the condition over at least a reach level or over the entire watershed system. The goal of the assessments is to better understand the processes so that stream restoration, bank stabilization, or a host of other river applications can be designed successfully. These methods are discussed briefly below.

Pfankuch developed a method to rate stream stability for mountain streams in the northwestern United States.[5] The methodology was developed for the purpose of planning various stream projects on second- to fourth-order streams. The user evaluates the condition of the stream by assessing 15 stability indicators. For each indicator, the user rates the stream reach as excellent, good, fair, or poor based on a set of qualitative descriptions for each category. Each indicator and rating is associated with a number of points. When the ratings have been completed, the points are added to yield a total score. The total score is then related to a subjective description for the overall stability of the stream as excellent, good, fair, or poor; the higher the number, the more unstable the stream. The following list of parameters is used to indicate stability: bank slope, mass wasting, debris jam potential, vegetative bank protection, channel capacity (includes width-to-depth ratios), bank rock content, channel obstructions, bank cutting, deposition, angularity of rocks on bed, brightness of rocks, consolidation of bed material, bed material size, scouring, and moss and algae present. Thorough descriptions of each parameter are provided. The use of several of these parameters to infer stability is questionable. For example, brightness is used to describe the polishing of rocks, presumably due to movement. If less than 5 percent of the bottom is "bright," then the brightness is rated as excellent; if 5 to 35 percent of the bottom is "brighter" than the rest, then it is rated good, and so on. Not only is this a difficult parameter to evaluate, but also it would be highly variable in meaning from one stream to the next. The use of channel capacity to assess stability is also very subjective and problematic. In this method, the rating is excellent if the width-to-depth ratio, w/y, is less than 7, the cross section is ample for present peak volumes, and out-of-bank floods are rare. This may not be an appropriate rating, since the combination of w/y is less than 7 and "rare" bank floods may indicate an incising, unstable channel.

HEC-20 is a manual for bridge owners and inspectors to assess channel stability and potential stability-related problems in the vicinity of bridges and culverts.[3] A suggested three-level approach covers: (1) geomorphic concepts and qualitative analysis; (2) hydrologic, hydraulic, and sediment transport concepts; and (3) mathematical or physical modeling studies. If the results of level 1 suggest that the channel may be unstable in either the vertical or lateral direction, then the user is guided to continue to level 2. Based on those results, the user may or may not be instructed to continue to level 3. Level 1 is the qualitative analysis of geomorphic conditions leading to instability. Therefore, the details of this level will be described here. In level 1, the user completes a six-step process to determine the lateral and vertical stability and the potential response of the channel to changes. Step 1 is the collection of geomorphic data, such as stream size, flow habit, bed material, valley setting, flood plain and levee description, incision, channel boundaries, bank vegetation, sinuosity, braiding, and bar development. Each of these indicators is described in HEC-20. Step 2 involves reviewing historic changes in land use. Step 3 requires an assessment of overall stability based on data collected in step 1 as well as additional factors, such as dam and reservoir location, head cuts, sediment load, bed material size, flow velocity, and stream power. In step 4, lateral stability is evaluated as highly unstable, moderately unstable, or stable based on bank slope, bank failure modes, bank material, vegetation, and historic channel migration. In step 5, vertical channel stability is evaluated based on historic gradation changes and site observations. Finally, in step 6, the channel response to changes in sediment discharge, flow rate, bed slope, and sediment size is predicted.

Johnson et al. developed a rapid stability assessment method based on geomorphic and hydraulic indicators.[1] This method has been included in the most recent revision of HEC-20. It can be used within HEC-20 as a method to provide a semiquantitative level 1 analysis and to determine whether it is necessary to conduct a more detailed level 2 analysis. Thirteen qualitative and quantitative stability indicators are rated, weighted, and summed to produce a stability rating for gravel bed channels. It was based largely on previous assessment methods.[5,6,7] The primary limitation of the method is that it was developed and tested only in the Piedmont and glaciated Appalachian Plateau regions of Maryland and Pennsylvania.

Mitchell[29] and Gordon et al.[30] describe qualitative reconnaissance type surveys to assess stability of streams in Victoria, Australia. Field evaluations at each site typically were completed in only a few hours due to the nature of the sampling. Much of the sampling was completed by comparing the reach conditions to a set of drawings to categorize bank shape, channel shapes, bed material, and types and shapes of bars. However, criteria such as bank stability and bed aggradation or degradation were omitted from the data collection process due to the inconsistency of that information. The parameters that were evaluated and rated as very poor, poor, moderate, good, and excellent with respect to stability were bed composition, proportion of pools and riffles, bank vegetation, verge (riparian) vegetation, cover for fish, average flow velocity, water depth, underwater vegetation, organic debris, and erosion/sedimentation. The ratings for each parameter were based on qualitative descriptions. An overall rating then was assigned to each site based on the ratings of the 10 variables listed. However, the method used to determine the overall rating was not discussed.

Based on previous work by Simon,[24] Simon and Downs[6] developed a method for assessing stability of channels that have been straightened. In this method, a field form is provided for data collection in a 1.5- to 2-hour period. The data then are summarized on a ranking sheet. For each category on the ranking sheet, a weight is assigned where the value of the weights was selected based on the authors' experience. A total rating is derived by summing the weighted data in each category. The higher the rating, the more unstable the channel is. Simon and Downs found that for streams in western Tennessee, a rating of 20 or more indicated an unstable channel that could threaten bridges and land adjacent to the channel.[6] The rating system provides a systematic method for evaluating stability; however, the final ratings cannot be compared to streams evaluated in other geomorphic, geologic, or physiographic regions. In addition, some of the parameters are very difficult to assess, particularly in the absence of a stream gage. For example, considerable weight is placed on identifying the stage of channel evolution. To properly assess this stage, it is necessary to determine whether the channel is in the process of widening, degrading, or aggrading. Simon and Hupp provide a good description of determining bed degradation based on gage data.[31] However, determining aggradation or degradation based on a gage analysis typically requires at least several years of stream gage data.[3] Simon and Downs also have included rating information for bridges in the reach; this information can be used for geomorphic instabilities in the vicinity of the bridge, but does not incorporate local instabilities, such as bridge scour.[6]

Thorne et al. expanded on the method developed by Simon and Downs by adding a quantitative segment based predominantly on hydraulic geometry analysis.[7] The ranking based on the Simon and Downs method provides a qualitative assessment, while comparing measured hydraulic geometry to that calculated from equations developed for stable channels provides a quantitative measure of stability. A set of hydraulic geometry equations was assembled for gravel bed rivers, and the use was demonstrated on an actual river. The observed width and depth of a stream reach were compared to the regime width and depth calculated from the hydraulic geometry equations. Significant differences can then be assumed to imply that the observed channel is either in regime, or it is not. Although this is a reasonable approach, hydraulic geometry equations must be used cautiously because they are derived empirically. In addition, Merigliano showed that hydraulic equations do not always reflect channel behavior because of variability in other important parameters, such as turbulence, sediment distribution, and velocity distribution, that are not included in the equations.[32] Therefore, while it may be useful to use hydraulic geometry equations as a check on stability, the equations imply a level of accuracy and applicability that may not be appropriate. The qualitative and quantitative information is then assembled into a one-page report that summarizes the state of the channel stability. Writing the summary requires a great deal of field experience, because it is necessary to draw inferences from the qualitative and quantitative data.

In an attempt to reduce the amount of time required for a full geomorphologic study, Fripp et al. developed a stream stability assessment technique based on a one-page field form.[33] The required data to be collected include a basic description of the reach; restoration potential (or needs); channel bed condition in terms of whether it is stable, aggrading, or degrading; grade and bank controls; debris jams present; bank cover; bank erodibility (in terms of low, medium, or high); riparian buffer width; channel bed material; and cross-sectional measurements. From the collected data, it is suggested that the assessor rate the channel stability as good, fair, bad, or very bad based on qualitative descriptions of the channel bed and bank. It is not clear how all of the data collected on the field form are used in making the assessment. The authors stress that the stability assessment be based on the stream condition, not on any structures crossing the stream.

Myers and Swanson[34,35] applied the method developed by Pfankuch[5] to assess and monitor stream channel stability for streams in northern Nevada. They correlated the stream stability ranking to the stream type according to the Rosgen classification scheme.[36] Myers and Swanson found that several of the stability indicators proposed by Pfankuch were not useful in the evaluation. Based on these findings, they deleted rock angularity from the rating procedure and separated the combined scour and deposition indicator into two individual indicators. They also made slight adjustments to the scoring procedure. In addition, they found that if the rating was combined with a stream classification, the underlying morphological processes could be inferred from the classification, which then could be used to indicate an appropriate engineering response to mitigate further stream instability.

Montgomery and MacDonald suggest a diagnostic approach in which the system and system variables are defined, observations are made to characterize the condition of the system, and an evaluation is made to assess the causal mechanisms producing the current condition.[20] Observations are based on characterizing both the valley bottom and the active channel according to a set of field indicators. Valley bottom indicators include the channel slope, confinement, entrenchment, riparian vegetation, and overbank deposits. Indicators for the active channel include the channel pattern, bank conditions, gravel bars, pool characteristics, and bed material.

Rosgen proposed a channel stability assessment method that is based on assessing stability for a stable reference reach, then assessing the departure from the stable conditions on an unstable reach of the same stream type.[37] The stability analysis consists of 10 steps that assess various components of stability. The steps include measuring or describing: (1) the condition or "state" categories (riparian vegetation, sediment deposition patterns, debris occurrence, meander patterns, stream size or order, flow regime, and alterations); (2) vertical stability in terms of the ratio of the lowest bank height in a cross section divided by the maximum bankfull depth; (3) lateral stability as a function of the meander width ratio and the bank erosion hazard index (BEHI); (4) channel pattern; (5) river profile and bed features; (6) width-to-depth ratio; (7) scour and fill potential in terms of critical shear stress; (8) channel stability rating using a modification of the Pfankuch[5] method; (9) sediment rating curves; and (10) stream type evolutionary scenarios. This is a very data-intensive assessment method and not one that bridge inspectors or hydraulic engineers will likely use, due to the time and expense of data collection. However, one of the more interesting components of this method is the procedure for step 8. Like the Pfankuch method, a rating of good, fair, or poor is obtained based on a numerical rating. However, Rosgen modified the method to account for differences across the 42 different stream types, so that each stream type has a separate definition for good, fair, and poor. For example, a rating of 60 would be considered poor in a B1 stream, fair in a C1 stream, and good in an F1 stream. Although the approach is interesting and has merit, the basis for the 42 separate rating schemes is not given.

USACE suggests a three-level stability analysis for the purpose of stream restoration design.[10,38] Level 1 is a geomorphic assessment, level 2 is a hydraulic geometry assessment, and level 3 is an analytical stability assessment that includes a sediment transport study. As part of the geomorphic assessment, USACE recommends collecting the following field data: watershed development and land use, flood plain characteristics, channel planform, and stream gradient; historical conditions; channel dimensions and slope; channel bed material; bank material and condition; bedforms, such as pools, riffles, and sedimentation; channel alterations and evidence of recovery; debris and bed and bank vegetation; and photographs. Indicators of channel degradation are given as terraces, perched channels or tributaries, head cuts and knickpoints, exposed pipe crossings, perched culvert outfalls, undercut bridge piers, exposed tree roots, leaning trees, narrow and deep channels, undercut banks on both sides of the channel, armored beds, and hydrophytic vegetation located high on the banks. Indicators of a stable channel include vegetated bars and banks, limited bank erosion, older bridges, culverts and outfalls with inverts at or near grade, no exposed pipeline crossings, and tributary mouths at or near existing main stem stream grade. Copeland et al. further suggest that spatial bias in assessing stability can be reduced by walking a distance well upstream and downstream of the project reach, while temporal bias can be reduced by revisiting the site at different times of year.[38] The USACE manual on assessing channel stability for flood control projects provides a detailed example of a quantitative stability analysis, based primarily on critical and design flow shear stresses.[10]

Annandale developed a two-level procedure to determine the risk of bridge failure that included river instability.[39,40] The first level is a hazard assessment and procedure for rating hazards. The hazard assessment is comprised of river instability, potential for morphological change, fluvial hydraulics in the immediate vicinity of the river crossing, and the structural integrity of the river crossing. Annandale provides four tables of values to assign for each of these factors. The values were based on river crossing failures in New Zealand, South Africa, and the United States. The hazard rating is the product of the four values. The table for assessing the hazard rating for the river stability factor is based on channel type from Schumm.[41] The values of the factors range from 1.00 for a straight, suspended load channel, to 3.162 for a braided, bed load channel. A second table provides ratings for the potential for morphological changes (degradation, bank erosion, and aggradation) due to extraneous factors. Annandale accounts for the location of the bridge with respect to stream meanders as a separate factor in his method. If the bridge is between meanders or on a tight bend, the factor value is increased. The hazard rating, based on the product of the four factors, is categorized as significant, moderate, or low.

Individual States have also developed protocols and methods for assessing stream stability. For example, the Vermont has assembled an extensive manual on stream stability assessment.[42] The State's method follows that of Pfankuch.[5] The manual includes a field form for bridges and culverts; however, it is primarily an inventory for habitat disruptions, rather than part of the stability assessment.

In addition to specific indicators listed for each method, Shields determined that factors in the watershed should be examined as part of assessing current and future channel stability.[43] Watershed characteristics include:
- Physical characteristics and the channel network. Shields suggests using multiple classification methods, such as Rosgen,[36] USACE,[38] Schumm,[41] and Harvey and Watson,[44] to classify these physical characteristics. Applying multiple methods is useful in that each method provides different results and information.
- Nature of existing and future hydrologic response and sediment yield. Water and sediment discharges are affected by urbanization, deforestation, mining, logging, and other disturbances. Changes will cause a response in the stream channel, possibly creating instability.
- Existing instability in the overall system and the causes. Depending on the problem, channel instabilities can move upstream or downstream, possibly moving into a project area and causing destabilization there.

Bank erosion can be categorized as either fluvial erosion or mass wasting (geotechnical). Thorne and Osman showed that for a given set of soil conditions, there is a combination of critical bank height and angle greater than which the bank will be unstable (see figure 4).[45] Although worthy in concept, determining the critical bank height and angle requires significant field observations and measurements. Factors that influence fluvial bank erosion include bank material, stream power, shear stress, secondary currents, local slope, bend morphology, vegetation, and bank moisture content.[46] Factors that influence mass wasting include bank height, angle, material, and moisture content.

Other researchers also have developed lists of parameters that indicate stability without defining how they can be used to assess overall channel stability. For example, Lewin et al. provided a list of indicators and indicated whether they affect lateral or vertical stability.[47] Most of these indicators are obvious and provide only the general direction of instability.

Many stream channel stability indicators are common to multiple assessment methods discussed above. These indicators are summarized in table 2. Characteristics of those indicators also are provided. Additional information is given in the references.

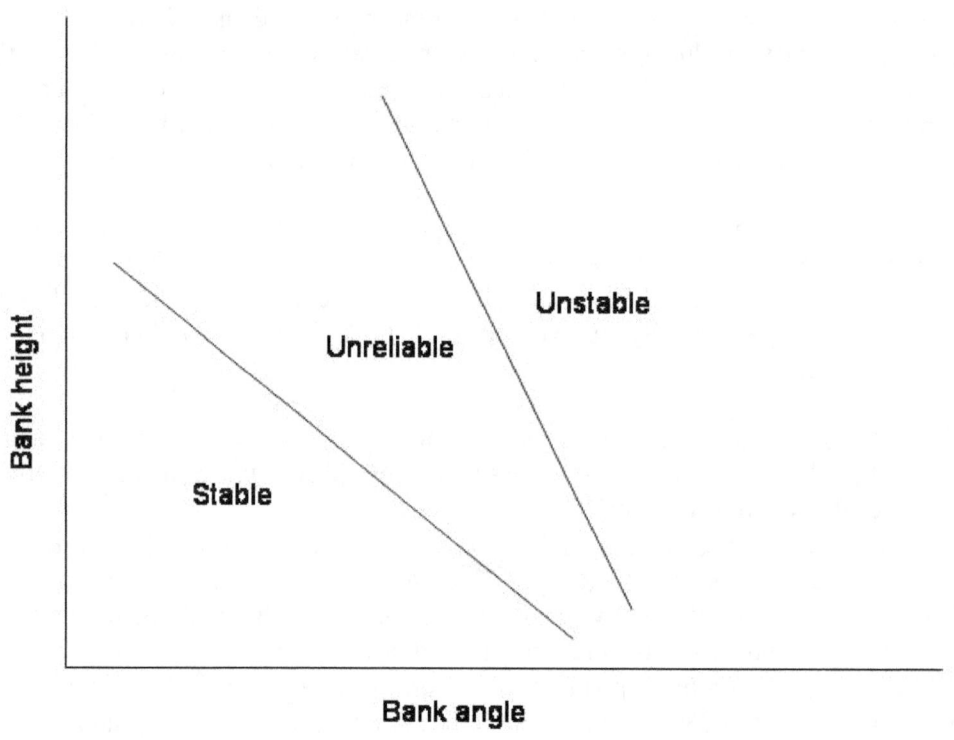

Figure 4. Critical bank height and angle.

Table 2. Summary of common indicators used in channel stability assessment methods.

Indicator	Characteristics	References
Boundary conditions	Changes in loads to the stream, including water and sediment inputs. Causes channel to adjust. Amount and type of adjustment primarily based on channel type.	17, 20
Flow habit and flashiness	Ephemeral, intermittent, or perennial; flashiness; stream order. Flashy urban or ephemeral streams tend to be unstable relative to nonflashy perennial streams. Large ephemeral streams more unstable than first order.	3, 10, 48
Valley setting, confinement	Confinement by narrow valley, rock walls, or revetments limits lateral migration and may promote downcutting.	3
Channel pattern and type	Indicates energy in the system, propensity for lateral migration, processes on larger scale.	3, 11, 20, 37, 40, 41, 48, 49, 50
Entrenchment, incision	Indicated by levees, terraces, low width-to-depth ratio, confinement, exposed infrastructure. Indicates processes that have been occurring.	2, 20
Bed material	Size, uniformity, and packing are typical measurements and observations that indicate movement, armoring, and energy of the system. Fraction of sand and gravel can also indicate sediment transport characteristics.[51,52]	5, 10, 11, 30, 48, 49, 50, 51, 52, 53, 54, 55
Bar development	Indicators of bar stability include the relative bar size, bar material, and vegetation. For streams on higher slopes (> about 0.02) and low width-to-depth ratios < 12, bars are not typically evident.	3, 20
Bed and bank obstructions	Obstructions include rock outcrops, cohesiveness, grade control, bridge bed paving, debris jams, dikes or vanes, and revetments. They can cause flow diversions that erode beds and banks.	2, 5, 20

Table 2. Summary of common indicators used in channel stability assessment methods, continued.

Indicator	Characteristics	References
Bank material	Layers/lenses, material size and sorting, and cohesiveness indicate the coherence of the material and resistance to erosion.	10, 56, 57, 58
Bank angle and height	Critical heights can be determined, greater than which the bank is susceptible to failure. Stable banks have low angles and heights.	5, 45, 57, 58
Bank and riparian vegetation	Woody vegetation helps maintain stability. The type, diversity, density, health, vertical orientation, and maturity of woody vegetation are indicators of stabilization effects. Rosgen includes root density.[37]	5, 7, 37
Bank cutting	The relative amount of bare banks and root exposure along a bank indicate stability.	5, 20
Mass wasting	Indicated by scalloping of banks, irregular channel width, tension cracks, slumping.	3, 5

STREAM CLASSIFICATION

A stream classification scheme is a method of classifying a stream according to a set of observations. Streams are usually classified for the purpose of communication; a description of a stream by a classification gives the reader or audience an immediate picture of the appearance and condition of that stream channel and possibly its relationship to the surrounding flood plain and other streams in the system. More recently, classification schemes also are being used as a basis for channel restoration designs. There are a variety of classification schemes; the choice of scheme is problem dependent. Niezgoda and Johnson list 27 classification schemes devised over time, beginning in 1899, and a brief description of each.[59]

Although the idea of classifying a stream may initially seem to be a trivial matter, the many complexities of stream configuration and planform often make classification difficult. In addition, classification schemes tend to force the stream into a category which may be useful for communicating the condition of a stream, but may be detrimental in that it may overlook certain unique or unusual characteristics of a stream. For example, few, if any, stream classification schemes include characteristics of streams in highly urbanized settings. Therefore, to use an available classification scheme, the urban stream is frequently forced into a particular classification. While this may communicate some characteristics of the stream, it ignores others.

One of the most commonly used and useful classification schemes is stream order. The order of a stream describes the relationship of the stream to all other streams in the watershed. Streams that have no tributaries flowing into them are ranked as number one, or first-order streams. A second-order stream is one that is formed by the junction of two first-order streams or by the junction of a first- and a second-order stream. This ranking scheme is continued for all channels within the drainage basin. Stream order increases in the downstream direction, and only one stream channel can have the highest ranking. This process of ranking is a rather simple matter for a small drainage basin, but can become very difficult for a large, complex basin. Various stream characteristics have been related to stream order. For example, channel slope and channel length can be related to stream order in a given basin. First-order streams typically are steeper and shorter than second-order or higher streams. Fourth-order streams typically are relatively large, wide, low-gradient streams. This information can be useful in determining various characteristics about drainage basins, particularly large basins where extensive data gathering is impractical.

Streams also can be classified according to other physical channel characteristics. These characteristics are qualitatively described by variables such as point bars, meanders and braiding, bank material, and valley slope. Brice and Blodgett developed a classification system for streams based on qualitative observations about the channel width, flow habit, flood plain, degree of sinuosity and braiding, development of point bars, bank material, and vegetative cover on the banks.[60] Rosgen developed a similar, but more extensive, stream classification scheme that has recently come into widespread use.[36,61] In this scheme, a stream is classified according to sinuosity, channel slope, bed material, entrenchment ratio, and width-to-depth ratio. The Rosgen scheme categorizes streams as six different types, A through G. Each type is associated with a range of slopes, width-to-depth ratios, sinuosities, and entrenchment. The streams are further subdivided according to the median size of the channel bed material. For example, a C5 stream is a low gradient, meandering stream with a high sinuosity and a sand bed.

23

Montgomery and Buffington developed a stream classification scheme that also has been used widely in recent years.[9] The stream type is based on the location within the watershed, the response to the sediment load, and several physical attributes. Table 3 provides the stream types and their characteristics. A primary advantage of using this method is that it is relatively simple and provides information on processes according to channel type. The channel types refer to natural, unmodified channels.

USACE developed a method for classifying streams based on location and processes within a watershed.[10] Figure 5 depicts the stream types that are based on location. In addition, there are categories for arroyos, underfit (glaciated) streams, regulated streams, deltas, and modified channels. Classifying a given stream channel is entirely based on the descriptions given in figure 5. The advantage of this method is that it includes engineered or modified channels, which are common across the United States.

Table 3. Montgomery and Buffington stream classification system.[9]

	Braided	Dune-Ripple	Pool-Riffle	Plane-Bed	Step-Pool	Cascade	Bedrock	Colluvial
Typical Bed Material	Variable	Sand	Gravel	Gravel, cobble	Cobble, boulder	Boulder	N/A	Variable
Bedform Pattern	Laterally oscillary	Multilayered	Laterally oscillary	None	Vertically oscillary	None	N/A	Variable
Reach Type	Response	Response	Response	Response	Transport	Transport	Transport	Source
Dominant Roughness Elements	Bedforms	Sinuosity, bedforms	Bedforms, grains, LWD, sinuosity, banks	Grains, banks	Bedforms, grains, large woody debris (LWD), banks	Grains, banks	Boundaries	Grains, LWD
Dominant Sediment Sources	Fluvial, banks failure, debris flow	Fluvial, bank failure, inactive channel	Fluvial, bank failure, inactive channel, debris flows	Fluvial, bank failure, debris flows	Fluvial, hill slope, debris flow	Fluvial, hill slope, debris flow	Fluvial, hill slope, debris flow	Hill slope, debris flow
Sediment Storage Elements	Overbank, bedforms	Overbank, bedforms, inactive channel	Overbank, bedforms, inactive channel	Overbank, inactive channel	Bedforms	Lee and Stoss sides of flow obstructions	N/A	Bed
Typical Slope	$S < 0.03$	$S < 0.001$	$0.001 < S < 0.02$	$0.01 < S < 0.03$	$0.03 < S < 0.08$	$0.08 < S < 0.30$	Variable	$S > 0.20$
Typical Confinement	Unconfined	Unconfined	Unconfined	Variable	Confined	Confined	Confined	Confined
Pool Spacing (Channel Widths)	Variable	5 to 7	5 to 7	None	1 to 4	<1	Variable	Variable

Mountain Torrents. High velocity streams on steep slopes with a drop-and-chute structure often achieved by obstacles such as large boulders or debris. These streams are subject to scour and degradation caused by flood events. Very steep slopes can lead to debris flows that produce substantial movement of boulders and gravels.

Alluvial Fans. Occur usually in arid and semiarid lands where a stream flowing through a stream valley enters a flat area. The coarse sediment carried by the stream deposits in a delta-like configuration characterized by multiple channels subject to shifting. The chief stability problem is caused by the unpredictability of the flow paths, which may cause erosion and deposition in unexpected places.

Braided Rivers. The main characteristic of these streams is a series of interlaced channels defined by bars and islands. Braided streams often occur in upper and middle zones of the watershed and usually involve gravel and cobbles, although braiding may also occur in sands. Scour and deposition often cause shifting of the main channel.

Arroyos. Present in arid and semiarid lands, these are streams that remain dry most of the time and carry flow only during flood events. Discharge and sediment transport can be substantial during flow episodes. Incising channels, width enlargement, and deposition are typical problems associated with arroyos.

Meandering Alluvial Rivers. These occur primarily in the middle and lower portion of the watershed. The planform of the stream is characterized by meanders that erode the streambank in the outer side of the bend and deposit material on the inner side. Meanders may migrate in the flood plain and can often become cutoff periodically when two bends advance toward each other and curvatures becomes severe. Cutoff meanders become isolated features called "oxbow lakes" that eventually fill with sediment. Traces of old meanders (scrolls) are easily distinguishable in aerial photographs. Measures that alter the supply of water or sediment have the potential to change cross sections, planforms, and gradients.

Modified Streams. This term generically encompasses those streams whose natural configuration has been severely modified by human intervention. These modifications include straightening, channelizing, enlargement, and base level changes caused by regulation of the receiving stream. Increased runoff from surrounding development also introduces modifications.

Regulated Streams. Regulation of tributaries by upstream reservoirs reduces flood flows and increases baseflow. These changes in the flow regime translate into reduced morphological activity. If regulation facilitates sediment deposition in the channel and vegetation growth, the stream cross section will be reduced. However, if the stream carries substantial sediment loads that become trapped in the reservoir, the stream may cause erosion downstream of the dam.

Deltas. These features occur on flat slopes of the lower portion of the stream where it empties into relatively quiescent water such as the ocean or a lake. Sediment deposition due to reduced velocity forces the river to split into distributaries whose base level rises as the delta progresses into the water body. Deltas also exhibit the formation of natural levees along the distributaries.

Underfit Streams. These are streams common in regions whose landscape formed as a result of glacial activity. Underfit streams occur in wide valleys formerly shaped and occupied by larger streams, usually the outlet to glacial lakes. Underfit streams are also found in abandoned riverbeds or channels downstream from reservoirs. Flat slopes, low velocities, and established vegetation make underfit streams generally stable.

Cohesive Channels. These are channels cut in cohesive materials such as marine clays, silted lakes, and glacial till plains. In marine deposits, these streams behave somewhat like meandering alluvial streams, although the meanders are flatter, wider, more uniform, and usually more stable. In glacial till, the planform tends to be irregular.

Figure 5. USACE[10] stream classification system.

PHYSIOGRAPHIC REGIONS

The United States can be divided into eight major physiographic regions based on geologic and geomorphologic characteristics. The major regions can be further subdivided into subregions, for a total of 25 physiographic regions. The boundaries of these regions are not strictly defined. Rather, the regions are approximately mapped to provide a description of major physiographic changes across the country. Figure 6 shows these regions based on a map created by Fenneman and Johnson.[62] The regions and subregions are:

- Region 1—Laurentian Highlands, including Superior Uplands.
- Regions 2–3—Atlantic Plain, including Coastal Plain and Continental Shelf.
- Regions 4–10—Appalachian Highlands, including Piedmont, Blue Ridge, Valley and Ridge, St. Lawrence, Appalachian Plateau, New England, Adirondack.
- Regions 11–13—Interior Plains, including Interior Low Plateau, Great Plains, Central Lowlands.
- Regions 14–15—Interior Highlands, including Ozark Plateaus, Ouachita Province.
- Regions 16–19—Rocky Mountain System, including southern Rocky Mountains, Wyoming basin, middle Rocky Mountains, northern Rocky Mountains.
- Regions 20–22—Intermontane Plateaus, including Columbia Plateau, Colorado Plateau, Basin and Range.
- Regions 23–25—Pacific Mountain System, including Cascade-Sierra Mountains, Pacific border, lower California.

Physiographic regions provide natural divisions by which to investigate stream processes and erosion issues broadly. Thornbury defined a physiographic unit as an area of land with similar or uniform topographic characteristics, including altitude, relief, and type of landforms, that are distinctly different from other physiographic units.[63] Dietz suggested that the erosion and sedimentation processes and the rates of those processes also define the topography and landforms.[64] Several publications describe the landforms and underlying geologic structure of each of the physiographic regions listed above.[63,65,66] The landforms result from the combination of the underlying geologic structure and the erosion processes on the surface. Thus, gross characteristics of the streams can be summarized for each of the physiographic Provinces. Within each of these Provinces, however, there is a range of stream types due to variability in valley slope, sediment supply, and water discharge. In addition, changes to the stream channels through engineering of channels (straightening, clearing, widening, dredging) and removal of riparian vegetation also have tremendous impacts on the form that the channel will take.

A number of studies supply evidence of the link between stream channel characteristics and physiographic region. Based on field observations and a detailed literature review, Graf characterized the physiographic Provinces and the stream types in each.[67] The recent activities in stream restoration have motivated a number of studies that attempt to characterize the morphologic characteristics of stream channels within a specific physiographic region. Most of these studies are focused on developing so-called regional equations that provide stream width and depth as a function of drainage area and/or bankfull discharge. These equations typically are developed within specific physiographic Provinces, thus providing additional evidence of common stream channel characteristics as a function of physiography. When comparing the

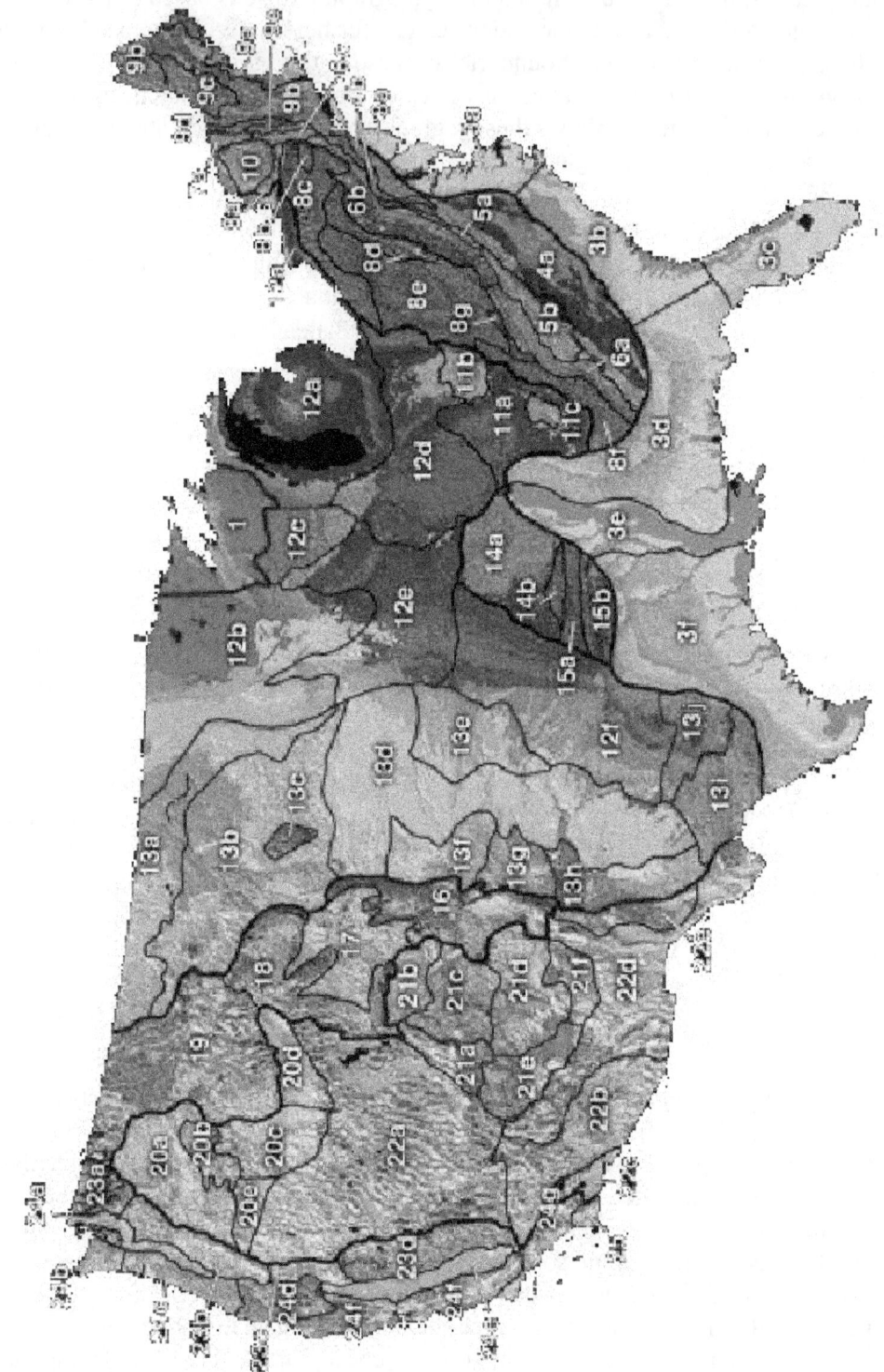

Figure 6. Physiographic map of the United States (after reference 66).

regional equations with other sites, it should be kept in mind that the data for the equations are almost always collected at stream gages, which means that the site is likely to be stable. Thus, a comparison of widths and depths obtained from these equations with those measured at other sites may indicate relative stability of the observed site. These studies and others are summarized below for streams in selected regions to illustrate the variation in streams from one region to another.

Appalachian Highlands. Streams in this region are generally meandering and perennial; however, the pattern is greatly influenced by slope, geology, and bed materials.[68] In the higher elevations, the geology is generally (although with exceptions) more resistant material. Hack showed that slopes in the Shenandoah Valley are much steeper (about seven times steeper) than those in the Martinsburg, shale areas, while streams in the carbonate rock areas had slopes in between those of the Shenandoah Valley and Martinsburg, shale areas.[69] However, Hack also showed that the slopes of channels across the Appalachian region could be predicted by $18(d/A_D)^{0.6}$[70], where A_D is the drainage area d is the median size of the bed material. Channel pattern in this region is also controlled by bedrock. The bedrock exposures and rough terrain tends to create waterfalls and fast-moving streams across the region.[71] Regional equation parameters, developed by the U.S. Geological Survey for the Appalachian Plateau of New York, are given in table 5.[72] The equations resulted in a width-to-depth ratio of $16.5A_D^{0.09}$, which can be approximated as 16.5, since the exponent of 0.09 is quite small and has little effect on the width-to-depth ratio. Piedmont streams are moderately sloped, controlled by bedrock outcroppings.[71] Bed material is primarily sand and gravel.[73] A number of studies have determined the hydraulic geometry of Piedmont streams. The resulting regional equation parameters are given in table 4. Variation within the resulting width-to-depth ratios are significant, ranging from about 5.04 to 12.53. The range can be attributed to variability in levels of urbanization, land use, location within the Piedmont, and differences in observer identification of bankfull elevations.

Coastal Plain. Streams are perennial and primarily meandering with limited reaches of braiding where channel slopes are higher and sediment loads are greater.[74] Oxbow lakes, back-swamps, and natural levees are common. Engineered dams and levees are also common to control flooding. Coastal Plain stream bottoms consist of more easily erodible material than the neighboring Appalachian Highlands.[71] Stream slopes are primarily gentle with bed material consisting of sand to fine gravel. Studies have been conducted in several eastern States to determine the hydraulic geometry of streams in the Coastal Plain region. Sweet and Geratz developed the set of regional equation parameters for North Carolina's Coastal Plain, given in table 4.[75] Prestegaard found slightly different equations for width and depth, also given in table 4.[73] She determined average bed materials to be in the sand and gravel ranges with an average median sediment size of 8 millimeters (mm). She also found that, in general, Coastal Plain streams were deeper and narrower than Piedmont streams. This can be seen by comparing the width-to-depth ratios for each region.

Great Plains and Central Lowlands. Streams in the Great Plains and Central Lowlands can be ephemeral, intermittent, or perennial. The Great Plains represent a depositional environment, deriving sediment from the mountains to the west. Deposited sediment is then reworked and moved through the channel system.[76] Channel morphology in the Midwestern United States has been affected by water development projects. Several well-known geomorphic studies in the Great Plains have related channel geometry and hydraulic geometry sediment characteristics and flow discharge.[80,81] Both the Central Lowlands and the Great Plains, as well as portions of the Interior Lowlands, are covered with thick deposits of loess.[82] Simon and Rinaldi conducted reconnaissance studies at streams throughout the Midwest and determined that the combination of easily erodible soils and extensive human disturbance has produced thousands of miles of highly unstable streams.[82]

Rocky Mountains. Local variations in geological setting and tectonics have significant control over fluvial processes and, thus, stream morphology.[83] Coarse-grained sediment is transferred from the mountains to the surrounding basins. Channels in the mountains are relatively steep and carry a coarse sediment load. In the adjacent basins, channels have lower gradients and transport finer grained materials that are derived, in part, from the coarser upstream load.

Basin and Range. This Province is primarily arid to semiarid. Thus, rivers in the Basin and Range tend to be ephemeral or intermittent. Alluvial fans are common in the Basin and Range. They develop when sediment transported along steep, mountain channels deposits on shallower slopes at the base of the mountains. Streams in alluvial fans are typically highly unstable in terms of lateral position. One of the most outstanding characteristics of rivers in the Basin and Range is that drainage across most of the Province is internal.[84] Castro and Jackson developed regional equation parameters, given in table 4, for the upper Basin and Range based on measurements at 22 stream channels.[79]

Pacific Coastal (California). The Pacific Coastal region in California is characterized by a wide variety of drainage types that include arroyos and alluvial fans.[85] Streams in both of these stream types tend to be unstable both laterally and vertically. Human alterations of stream channels in this region are widespread and have changed the erosion and depositional patterns. Streams may be ephemeral, intermittent, or perennial. Castro and Jackson developed regional equation parameters for this region, given in table 4.[79]

Table 4. Regional equation parameters for selected physiographic regions in the United States.

Reference	Region	a	b	c	d	f	g	i	j	w/y (Q)	w/y (A_D)
77	Piedmont	–	–	–	–	5.43	0.33	0.54	0.33	N/A	10.06
78	Piedmont	1.46	0.52	0.19	0.42	14.78	0.39	1.18	0.34	7.68	12.53
73	Piedmont	1.36	0.52	0.27	0.37	21.5	0.25	2.09	0.14	5.04	10.29
72	Appalachian Plateau	–	–	–	–	13.19	0.46	0.80	0.37	N/A	16.50
75	Coastal Plain	1.06	0.50	0.11	0.48	9.64	0.38	0.98	0.36	9.64	9.84
73	Coastal Plain	3.15	0.39	0.51	0.28	11.3	0.35	1.93	0.12	6.18	$5.85A_D^{0.23}$
79	Basin and Range	0.96	0.60	0.36	0.31	3.27	0.51	0.79	0.24	$2.67Q^{0.3}$	$4.14A_D^{0.27}$
79	Pacific Coastal	2.37	0.5	0.15	0.45	12.39	0.43	0.66	0.39	15.8	18.77
42	New England	2.65	0.47	0.62	0.23	10.18	0.50	1.22	0.25	$4.08Q^{0.23}$	$8.35A_D^{0.25}$

$w = aQ^b$, $y = cQ^d$, $w = fA_D^g$, $y = iA_D^j$; w/y (width-to-depth) ratios are approximated as constants for w/y with very small exponents for either Q or A_D. The variables a-j are regression coefficients

3. FIELD OBSERVATIONS

Numerous stream-bridge intersections were observed across the United States to develop and test the stability assessment method. The streams were to reflect a broad range of stream types and physiographic regions; thus, 57 site visits were conducted in 13 physiographic regions and subregions, including Pacific Coastal, Basin and Range, Trans Pecos, southern Rocky Mountains, Great Plains, Central Lowlands, Interior Lowlands, Ozark-Ouachita Plateau, Appalachian Plateau, Valley and Ridge, Piedmont regions, and Atlantic Coastal Plain.

In addition to collecting observations at streams covering a variety of erosion issues, sizes, and physiographic regions, the following criteria also were used in selecting appropriate sites:
- All channels were alluvial or partially alluvial (occasional rock outcrops were acceptable).
- Engineered (straightened or widened) channels were included, although manmade canals were not.
- The streams had to be wade-able or partly wade-able.
- The streams and bridges had to be safely accessible.
- A reasonable level of personal physical safety had to be satisfied.
- The streams were located within a reasonable distance of the travel corridor.

The data for each of the streams are summarized in tables 5–7. Streams that are named N# in table 5 are unnamed on topographic maps. Table 5 provides the locations and global positioning system (GPS) coordinates of the bridges, the physiographic Province, land use, and stream classification. Each of the channels was classified according to the Montgomery-Buffington scheme. The Montgomery-Buffington scheme does not include engineered or altered channels. However, this method is still useful as a basic descriptor of the primary processes in the stream (e.g., transport versus response) and is, therefore, included. To include altered streams, the USACE method[10] and a simple observation of channel pattern (based on both field observation and aerial photos) also were used to classify or categorize the stream types. The resulting stream type is provided as a combination of these methods in table 5. The Rosgen classification method was not considered because it is unnecessarily data intensive for the purposes of assessing channel stability in the vicinity of a bridge. Table 6 provides the bed and bar material, the percent of sand (F_s), and any controls observed in the banks or on the bed. Table 7 provides observations made on the banks, including vegetation, bank material, bank height, and any erosion characteristics. In the next section, observations made in each of the physiographic regions are described.

Table 5. River data summary.

River	Map Location	GPS Location		Physiographic Province	M-B/USACE Class*	Land Use	Channel Pattern
		N. Latitude (deg)	W. Longitude (deg)				
1. Saline	U.S. Rt. 183 22.5 kilometers (km) N of Hays, KS	39.0973	99.3055	Great Plains	D/MA	Cultivated	Meandering
2. S. Fork Solomon R.	U.S. Rt. 283 1.6 km S of Hill City, KS	39.3506	50.8457	Great Plains	D/MA-BR	Cultivated	Meandering to braided
3. N. Rush Cr.	State Route (S.R.) 71, S of Limon, CO	39.0609	103.7035	Great Plains	D/MA	Managed/grass	Meandering
4. Arkansas R.	S.R. 291, N of Salida, CO	38.6127	106.0618	Rocky Mtn.	R/MA	Natural/ cultivated	Meandering
5. Tomichi Cr.	S.R. 114, 12.9 km E of Gunnison, CO	38.5202	106.7852	Rocky Mtn.	R/MA,MO	Cattle pasture	Meandering
6. Murietta Cr.	Main St., Temecula, CA	33.4924	117.1499	Pacific Coastal	D/MA-BR, MO	Suburban	Meandering
7. Jacalitos Cr.	Jayne Ave., 9.7 km E of U.S. Interstate (I)-5 exit to Coalinga, CA	36.1369	120.2771	Pacific Coastal	D/MA-BR	Cattle grazing	Meandering to braided
8. Dry Cr.	Dry Creek Road, CA	38.4114	122.4513	Pacific Coastal	C/MT- MA	Wooded	Irregular
9. Dutch Bill Cr.	Bohemian Hwy., 9.7 km E of Oakville, Napa Valley, CA	38.4239	122.9569	Pacific Coastal	R-C/MA-MT	Wooded	Irregular
10. Buena Vista Cr.	S.R. 58 1.6 km E of Buttonwillow, CA	35.3995	119.5321	Pacific Coastal	B/MA-BR	Cattle grazing	Meandering to braided
11. Mojave R.	1st Ave., Barstow, CA	N/A	N/A	Basin and Range	B/BR	Rural/industrial	Braided

*C = cascade, S = step pool, P = plane bed, R = pool-riffle, D = dune-ripple, B = braided, MT = mountain torrent, MA = meandering, MO = modified, S.R = State Route, Cr. = Creek, R. = River

Table 5. River data summary, continued.

River	Map Location	GPS Location		Physiographic Province	M-B/USACE Class*	Land Use	Channel Pattern
		N. Latitude (deg)	W. Longitude (deg)				
12. Rt. 66 Wash	Rt. 66, E of Ludlow, CA	34.7160	116.1053	Basin and Range	B/AR	Natural	Arroyo/braided
13. Sacramento Wash	S.R. 68, 27.4 km E of Lauchlin, AZ	35.2250	114.2800	Basin and Range	R-B/MA-BR	Rural/mining	Meandering to braided
14. Rio San Jose	S.R. 6, 1.6 km S of I-40, Exit 126, NM	34.9675	107.1749	Trans Pecos	R/MA,MO	Grazed/natural	Meandering, channelized
15. Rio Puerco	S.R. 6, 24.2 km E of Los Lunas, NM	34.7966	106.9905	Trans Pecos	R/MA,MO	Grazed/natural	Meandering, channelized
16. W. Elk Cr.	E. Third St. (Rt. 66), W of Elk City, OK	35.4119	99.4522	Central Plains	R/MA,MO	Suburban, cattle	Meandering, channelized
17. Beaver Cr.	U.S. Rt. 183, ~1.6 km N of Arapahoe, OK	35.5941	98.9602	Central Plains	R/MA,MO	Agricultural, cattle	Meandering
18. Brush Creek	U.S. Rt. 62, 1.6 km E of Jacktown, OK	35.5068	96.9862	Central Plains	R/MA	Agricultural, grazed	Meandering
19. Unnamed	U.S. Rt. 62, 1.6 km E of Boley, OK	35.4862	96.4594	Central Plains	R/MA	Natural, grazed, rural	Meandering
20. Little Skin Cr.	U.S. Rt. 64, 1.6 km W of Muldrow, OK	35.3982	94.6211	Ozark-Ouachita Highlands	R/MA	Natural, rural	Meandering
21. Unnamed	U.S. Rt. 64, at Dyer, AR	35.4964	94.1363	Ozark-Ouachita Highlands	R/MA, MO	Rural	Meandering, straightened
22. Little Cypress Cr.	S.R. 59, 1.6 km S of I-40, Exit 35	35.3402	89.5018	Coastal Plain	R/MA	Agricultural, rural	Meandering

*C = cascade, S = step pool, P = plane bed, R = pool-riffle, D = dune-ripple, B = braided, MT = mountain torrent, MA = meandering, MO = modified, S.R = State Route, Cr. = Creek, R. = River

35

Table 5. River data summary, continued.

| River | Map Location | GPS Location | | Physiographic Province | M-B/USACE Class* | Land Use | Channel Pattern |
		N. Latitude (deg)	W. Longitude (deg)				
23. Unnamed	U.S. Rt. 70, 10.5 km W of Jackson, TN	35.6148	88.9909	Coastal Plain	R/MA	Agricultural, rural	Meandering
24. Unnamed	U.S. Rt. 79, 3.2 km SE of Milan, TN	35.8910	88.8105	Coastal Plain	R/MA	Agricultural, rural	Meandering
25. Honey Run	S.R. 76, 0.8 km E of White House, TN	36.4779	86.6397	Interior Low Plateau	P/MA	Rural	Slightly meandering
26. South Fork	S.R. 84 just NE of intersection with Rt. 357	37.5430	85.7685	Interior Low Plateau	P/MA	Agricultural, rural	Meandering
27. East Fork	S.R. 55 at intersection with U.S. Rt. 62, 3.2 km S of Bloomfield, KY	37.8816	85.3029	Interior Low Plateau	P/MA	Agricultural, rural	Meandering
28. Unnamed	U.S. Rt. 60 3.2 km E of I-40, Exit 101, ~ 9.7 km W of Sterling, KY	38.0443	83.9933	Interior Low Plateau	P/MA	Rural, grazed	Meandering
29. McKnown Cr.	U.S. Rt. 119 N at Robinson Rd., 2.4 km S of Walton, WV	38.5934	81.3792	Appalachian Plateau	P/MA	Rural	Meandering
30. Wolf Run	U.S. Rt. 199 N, 0.8 km S of Gandeville, WV	38.6896	81.3886	Appalachian Plateau	R/MA	Rural	Meandering
31. Alligator Cr.	S.R. 765, S of Punta Gorda, just S of US Rt. 41	26.8884	82.0213	Coastal Plain	D/MA	Suburban	Meandering
32. Peace R.	S.R. 70, 1.6 km W of Arcadia, Florida	27.2213	81.8766	Coastal Plain	D/MA	Suburban	Meandering

*C = cascade, S = step pool, P = plane bed, R = pool-riffle, B = braided, MT = mountain torrent, MA = meandering, MO = modified, S.R = State Route, Cr. = Creek, R. = River

Table 5. River data summary, continued.

River	Map Location	GPS Location		Physiographic Province	M-B/USACE Class*	Land Use	Channel Pattern
		N. Latitude (deg)	W. Longitude (deg)				
33. Blackrock Run	Stringtown Road off S.R. 25, N of Butler, MD	39.5437	76.7331	Piedmont	R/MA	Suburban, agricultural, natural	Meandering
34. Indian Run	Benson Mill Rd, off S.R. 25, N of Butler, MD	39.5691	76.7445	Piedmont	R/MA	Natural, agricultural, rural	Meandering
35. Middle Patuxent R.	S.R. 108, N of Columbia, MD	39.2290	76.9173	Piedmont	R/MA	Natural, suburban	Meandering
36. Hammond Branch	Stephens Rd., 2.4 km N of Laurel, MD	39.1318	76.8449	Coastal Plain	R/MA	Suburban, agricultural	Meandering
37. Atherton Tributary	Seneca Dr., Columbia, MD	39.1871	76.8629	Piedmont	R/MA	Suburban	Meandering
38. Stocketts Run	Sands Rd., 4.8 km SW of Davidsonville, MD	38.8831	76.6638	Coastal Plain	R/MA	Natural, rural	Meandering
39. Mill Stream Branch	S.R. 213, just S of Centreville, MD	39.0401	76.0722	Coastal Plain	R→D/MA, MO	Agricultural, rural	Meandering
40. Kent County Tributary	S.R. 446 (Broadneck Rd.), SW of Chestertown, MD	39.2039	76.1235	Coastal Plain	R→D/MA	Agricultural	Meandering
41. Morgan Creek	Kennedyville Rd., 1.6 km E of Kennedyville, MD	39.2969	75.9845	Coastal Plain	D/MA	Agricultural	Meandering
42. Little Elk Cr.	Little Elk Rd, 1.2 km N of PA-MD line, 2.4 km S of Hickory Hill, PA	39.7271	75.9078	Piedmont	P→R/ MA	Agricultural, rural	Meandering

*C = cascade, S = step pool, P = plane bed, R = pool-riffle, D = dune-ripple, B = braided, MT = mountain torrent, MA = meandering, MO = modified, S.R = State Route, Cr. = Creek, R. = River

Table 5. River data summary, continued.

River	Map Location	GPS Location		Physiographic Province	M-B/USACE Class*	Land Use	Channel Pattern
		N. Latitude (deg)	W. Longitude (deg)				
43. Big Beaver Cr.	Kuntz Mill Rd., off U.S. Rt. 222, 6.4 km N of Quarryville, PA	39.9411	76.2204	Piedmont	R/MA	Agricultural	Meandering
45. Roaring Run	PA Rt. 445, E of Nittany, PA	40.9810	77.5442	Ridge and Valley	S/MT	Natural	Step-pool
44. Buffalo Run	Fillmore Rd. (S.R. 3008) near State College, PA	40.8595	77.8769	Ridge and Valley	P/MA, MO	Agricultural/ rural	Meandering/ straightened
46. Potter Run	S.R. 144 at Potters Mills, PA	40.8013	77.6257	Ridge and Valley	S/MT	Natural/rural	Step-pool to meandering
47. Bentley Cr.	S.R. 4013 (Berwick Turnpike), about 1.6 km S of Bentley Creek, PA	N/A	N/A	Glaciated Appalachian Plateau	R/MA, MO	Rural residential	Meandering to braided
48. N 48	S.R. 58, 1.6 km W of Allegheny R., PA, S of I-80,	41.1340	79.6944	Appalachian Plateau	P, M/MT	Natural/rural	Plane bed, meandering
49. Reids Run	S.R. 68, S of I-80 at Reidsburg, PA	41.1469	79.4020	Appalachian Plateau	P, M/MT	Natural, agricultural, rural	Plane bed, meandering
50. Piney Creek	S.R. 66, 0.8 km S of Limestone, PA	41.1281	79.3277	Appalachian Plateau	P, M	Rural, agricultural	Plane bed, meandering
51. Little Sandy Cr.	S.R. 3011 at East Branch Station, PA	41.0327	79.0509	Appalachian Plateau	P, M	Agricultural, rural	Meandering
52. Trout Run	S.R. 310, 1.6 km S of Reynoldsville, PA	41.0787	79.9016	Appalachian Plateau	P, M	Natural, agricultural, rural	Meandering

*C = cascade, S = step pool, P = plane bed, R = pool-riffle, D = dune-ripple, B = braided, MT = mountain torrent, MA = meandering, MO = modified, S.R = State Route, Cr. = Creek, R. = River

Table 5. River data summary, continued.

River	Map Location	GPS Location		Physiographic Province	M-B/USACE Class*	Land Use	Channel Pattern
		N. Latitude (deg)	W. Longitude (deg)				
53. Pootatuck R.	Walnut Tree Hill Rd. near Sandy Hook, CT	41.4376	73.2702	New England	P-R, MA	Suburban, natural	Meandering
54. Mill R.	Judd Rd., N of Easton Reservoir, CT	41.3017	73.2760	New England	P, MA, MO	Natural	Meandering, straightened
55. Aspetuck R.	Silver Hill Rd. at Easton, CT	41.2596	73.3249	New England	P, MA, MO	Natural, rural	Meandering, straightened
56. W. Br. Saugatuck R.	Stonebridge Rd. at Wilton, CT	41.1949	73.3875	New England	P, MA	Natural, suburban	Meandering
57. Mianus R.	June Rd., N of Merritt Pkwy, Stamford, CT	41.1048	73.5867	New England	P, MO	Natural, suburban	Meandering, straightened

*C = cascade, S = step pool, P = plane bed, R = pool-riffle, D = dune-ripple, B = braided, MT = mountain torrent, MA = meandering, MO = modified, S.R = State Route, Cr. = Creek, R. = River

Table 6. River channel data.

River	w/y	Bed Controls	Bank Controls	Fs (%)	Bed Material	Bar Type	Bar Material	Bar Width	Bar Vegetation
1. Saline	20	None	Abutments	80	Sand	Alternate	Sand	1/2 W	Grasses
2. S. Fork Solomon R.	17	None	Abutments	100	Sand	Alternate	Sand	1/3 W	None/grass
3. N. Rush Cr.	15	None	None	80	Sand	None	None	N/A	N/A
4. Arkansas R.	14	Gravel and cobble armor	Boulders, gravel armor, bridge abutments	30	Cobbles	Point bars, midchannel	Very coarse gravel-cobbles	1/5 W	Minimal
5. Tomichi Cr.	19	None	Right bank riprapped	50	Very fine gravel	Midchannel (d/s only)	Unknown	<1/5 W	Shrubs
6. Murietta Cr.	9	None	None	100	Sand	Alternate	Sand	1/2 W	Heavily vegetated
7. Jacalitos Cr.	23	None	Riprap at both bridge abutments	90	Sand	Irregular/combo	Coarse sand	>1/6, collectively very wide	Minimal
8. Dry Cr.	11	Boulders and occasional bedrock	Occasional bedrock and abutments	10 u/s	Cobbles	None u/s; Alternate/irregular d/s	Sand	1/2–2/3 W	None
9. Dutch Bill Cr.	10	None u/s	Bridge abutments	50 u/s	Fine gravel	Point bars	Gravel	2/3 W	None
10. Buena Vista Cr.	7.5–15	None	None	100	Sand	Irregular	Sand	1/2 W	None
11. Mojave R.	6.7	None	Piers, abutments	100 (fine)	Sand	Braided	Sand	Wide	None

Fs = portion of sand, u/s = upstream, d/s = downstream, W = width, R. = River, Cr. = Creek

Table 6. River channel data, continued.

River	w/y	Bed Controls	Bank Controls	Fs (%)	Bed Material	Bar Type	Bar Material	Bar Width	Bar Vegetation
12. Rt. 66 Wash	10	Boulders	Boulders, bridge protection, abutments	70	Sand, gravel	Irregular	Sand, gravel	Moderate	None
13. Sacramento Wash	20	None	Bridge protection, piers, abutments	75	Clay, silt, sand, gravel	Irregular	Sand and gravel	Moderate	None
14. Rio San Jose	5	None	Riprap, piers, abutments	70	Clay, silt, sand	None	N/A	N/A	N/A
15. Rio Puerco	16	None	Clay/silt cliffs, piers, bank stabilization	20	Silt, clay, sand	Alternate, point	Silt, clay	Wide	Grasses
16. W. Elk Cr.	9	Water line, debris	None	100 (silt)	Silt	None	N/A	N/A	N/A
17. Beaver Cr.	< 4	None	Bridge protection, soil blocks	100 (silt)	Clay, silt	Irregular	Clay/silt	Moderate	None
18. Brush Creek	5	Boulders from riprap	Bridge protection, bank stabilization	100 (silt, sand)	Clay, silt, sand	Irregular	Sand, silt, little gravel	None	Narrow
19. Unnamed	15	Boulders	Boulders, bridge protection, cliffs	100	Silt, sand	Irregular	Very fine sand	Wide	None, grasses

Fs = portion of sand, u/s = upstream, d/s = downstream, W = width, R. = River, Cr. = Creek

41

Table 6. River channel data, continued.

River	w/y	Bed Controls	Bank Controls	Fs (%)	Bed Material	Bar Type	Bar Material	Bar Width	Bar Vegetation
20. Little Skin Cr.	12	Beaver dams, debris	Bedrock, bridge protection	40	Sand, gravel	Alternate bars	Gravel	Grasses, shrubs	Moderate
21. Unnamed	5	Debris	Bars, debris	80	Silt, sand, gravel	Only under bridge	Sand	N/A	N/A
22. Little Cypress Cr.	9	Few boulders	Bridge abutments and pier	80	Sand w/ silt, gravel	Irregular	Sand	Narrow	None
23. Unnamed	9	Bridge protection	Debris, bridge protection	100	Sand	Irregular	Sand	Wide	None
24. Unnamed	9	Grade control	Bridge protection, piers	100	Sand	Irregular d/s only (u/s pooled)	Sand	Wide	None
25. Honey Run	12	Outcrop	Piers, bars	30	Gravel	Alternate	Gravel	Narrow	Grass/none
26. South Fork	9	None	Piers	30	Gravel	None	N/A	N/A	N/A
27. East Fork	9	Boulders, gravel armor	Bridge protection	50	Sand, gravel	None	N/A	N/A	N/A
28. Unnamed	12	Occasional bedrock, boulders	Occasional bedrock, abutments	20	Sand, gravel, cobbles	None	N/A	N/A	N/A
29. McKown Cr.	6	Gravel armor	Bridge protection	10	Gravel	None	N/A	N/A	N/A

Fs = portion of sand, u/s = upstream, d/s = downstream, W = width, R. = River, Cr. = Creek

42

Table 6. River channel data, continued.

River	w/y	Bed Controls	Bank Controls	Fs (%)	Bed Material	Bar Type	Bar Material	Bar Width	Bar Vegetation
30. Wolf Run	5–8	Gravel armor, occasional bedrock, grade control at bridge	Clumps of failed bank material	10	Gravel	Irregular (mostly from failure material)	Gravel	Moderate	None
31. Alligator Cr.	24	None	Confined left bank concrete wall or rock lined	18	Silt	None	N/A	N/A	N/A
32. Peace R.	20	None	Gravel armor (bank protection), bridge riprap, debris	100	Very fine sand	Irregular/ combination	Very fine sand	Wide, 1/2 W	Grasses, reeds, trees
33. Blackrock Run	5.5–17	Bedrock	Bedrock, abutments	65	Sand, gravel	Irregular	Sand, gravel	Narrow	None
34. Indian Run	5	Debris	Bridge riprap, abutments, fiber logs, debris	50	Sand, gravel, cobbles	Irregular	Sand, gravel	Wide	Grasses
35. Middle Patuxent R.	16	Gravel armor	Bedrock, gabions, riprap at bridge	20	Sand, gravel	Mid, point	Sand, gravel	Narrow	None
36. Hammond Branch	7	Boulders	Bridge protection, abutments, riprap	60	Sand, gravel	None	N/A	N/A	N/A

Fs = portion of sand, u/s = upstream, d/s = downstream, W = width, R. = River, Cr. = Creek

Table 6. River channel data, continued.

River	w/y	Bed Controls	Bank Controls	Fs (%)	Bed Material	Bar Type	Bar Material	Bar Width	Bar Vegetation
37. Atherton Tributary	16	Bedrock, boulders	Boulders, abutments	40	Sand, gravel, boulders	Irregular	Sand	Narrow	None
38. Stocketts Run	16	None	Abutments	50	Sand, gravel	Point	Sand, gravel	Moderate	None
39. Mill Stream Branch	13–16	None	Abutments, debris	100	Sand	Alternate	Sand	Narrow	None
40. Kent County Tributary	3	None	Abutments, concrete slabs	80	Silt, sand, gravel	Irregular	Silt, sand	Wide	Grasses
41. Morgan Cr.	10	None	Riprap, abutments	80	Silt, sand	None	N/A	N/A	N/A
42. Little Elk Cr.	20	Boulders	Bedrock, boulders, bridge protection, abutments	20	Sand, gravel, cobbles, boulders	Point	Gravel, cobbles	Narrow	None/grasses
43. Big Beaver Cr.	20–29	Gravel armor (d/s)	None	20	Silt, sand, gravel, cobbles	Midchannel (u/s), point (d/s)	Silt (u/s), sand, gravel (d/s)	Wide	None
44. Buffalo Run	10	Gravel armor, u/s weir	Bridge protection	30	Sand, gravel	None	N/A	N/A	N/A
45. Roaring Run	20	Boulders, riprap from failed bridge protection	Bridge protection, bank stabilization	10	Gravel, cobbles, boulders	Alternate (d/s only; u/s step-pool)	Gravel, cobbles	Moderate	None

Fs = portion of sand, u/s = upstream, d/s = downstream, W = width, R. = River, Cr. = Creek

Table 6. River channel data, continued.

River	w/y	Bed Controls	Bank Controls	Fs (%)	Bed Material	Bar Type	Bar Material	Bar Width	Bar Vegetation
46. Potter Run	15	Gravel/cobble armor	Bridge abutments, bank stabilization	30	Sand, gravel, cobbles	None	N/A	N/A	N/A
47. Bentley Cr.	32	None	Abutments	60	Gravel	Point, midchannel	Gravel	1/4–2/3 W	None
48. N 48	15	Gravel/cobble armor	Boulders, abutments, bank stabilization	10	Gravel, cobbles	None	N/A	N/A	N/A
49. Reids Run	15	Gravel/cobble armor, boulders	Boulders, abutments	10	Gravel, cobbles, boulders	None	N/A	N/A	N/A
50. Piney Creek	10	Gravel armor	Bridge protection, abutments, bank stabilization	< 20	Gravel, cobbles	Point bar	Gravel	Moderate	None
51. Little Sandy Creek	18	Gravel armor	Abutments	< 20	Gravel, cobbles, boulders	Midchannel	Gravel	Grasses	Narrow
52. Trout Run	10	Gravel armor	Bridge protection, abutments	30	Gravel, cobbles	None	N/A	N/A	N/A
53. Pootatuck R.	28	Bedrock, boulders, gravel armor	Boulders, abutments	10	Gravel, cobbles, boulders	Irregular	Gravel, cobbles	Narrow	None

Fs = portion of sand, u/s = upstream, d/s = downstream, W = width, R. = River, Cr. = Creek

Table 6. River channel data, continued.

River	w/y	Bed Controls	Bank Controls	Fs (%)	Bed Material	Bar Type	Bar Material	Bar Width	Bar Vegetation
54. Mill R.	13	Gravel armor, beaver dam d/s	Abutments	50	Sand, gravel	Alternate d/s	Sand, gravel	Narrow	None
55. Aspetuck R.	19	Gravel riffles	Boulders, bridge protection, bank stabilization	60	Sand, gravel	Alternate d/s (minor)	Sand	Narrow	None
56. W. Br. Saugatuck R.	45	Boulders	Boulders, abutments, debris jams, islands	40	Sand, gravel, cobbles	Midchannel, islands	Sand, gravel, cobbles	Moderate	Shrubs, trees
57. Mianus R.	13	Gravel armor (d/s of dam)	Boulders, abutments, bank stabilization	40	Gravel, cobbles	None	N/A	N/A	N/A

Fs = portion of sand, u/s = upstream, d/s = downstream, W = width, R. = River, Cr. = Creek

Table 7. River bank data.

River	Material	Bank Angle (deg)	Bank Height (m)	Vegetation	Erosion Location	Bank Failure Locations	Exposed or Bare Banks
1. Saline R.	Silty clay loam	30–60	0.9	Grass, deciduous trees; single dense and continuous band; healthy, vertically oriented	None	None	Minimal
2. S. Fork Solomon R.	Sand	30–60	0.6	Grass, reeds; no trees on channel banks; but in flood plain	Opposite and behind bars; adjacent to structures	No mass wasting	Frequent, but no cohesion
3. N. Rush Cr.	Loam/sand	60–80	0.9	Grasses	General fluvial; outside meander bends	Outside meanders	Outside meanders high, vertical
4. Arkansas R.	Sand, gravel, cobbles, silt	30–60	1.8	Grass, shrubs, deciduous and coniferous trees, healthy, vertical	Along straight reaches	Minor	Minor
5. Tomichi Cr.	Silty clay	Vertical	0.9	Grass, few deciduous trees (riparian, not bank)	Minor	None	None
6. Murietta Cr.	Sandy silty clay	60–80	1.5–2.4	Reeds, sparse deciduous trees well back from bank in flood plain, vertical	General, along channel banks	General; bank slides visible	Frequent, but sand so collapse
7. Jacalitos Cr.	Sand, gravel	30–60 (Vertical at outside bends)	0.9	Grass, deciduous trees	General fluvial, outside meanders	Outside meanders, recent slides	Occasional
8. Dry Cr.	Silty clay	Vertical	0.6–4.6	Dense deciduous trees; healthy, diverse, some leaning 30°	Outside meander; upstream of structure	General where too steep	Minor
9. Dutch Bill Cr.	Silt	Vertical	0.9–6.1 (incised)	Deciduous trees, some in channel (lateral migration)	Fluvial outside meander	Minimal	Some u/s of right abutment

LB = left bank, RB = right bank, u/s = upstream, d/s = downstream, R. = River, Cr. = Creek

Table 7. River bank data, continued.

River	Material	Bank Angle (deg)	Bank Height (m)	Vegetation	Erosion Location	Bank Failure Locations	Exposed or Bare Banks
10. Buena Vista Cr.	Sand/silt	60–90	0.6	Sparse shrubs	General, irregular	Slides in sand	Frequent
11. Mojave R.	Sand	60–90	1.8	None	General	Continual slides	Continual
12. Rt. 66 Wash	Clay, silt, sand	70–90	0.6–1.8	None	General	Frequent	Continual
13. Sacramento Wash	Clay, silt, sand, gravel	60–90	0.9	Few shrubs	General	Frequent slides, slumps	Continual
14. Rio San Jose	Clay, silt, sand	60–90	1.8	Desert shrubs	Outside meanders, general	Outside meanders	Continual
15. Rio Puerco	Clay, silt	80–90	1.5	Grasses, shrubs (desert)	Outside meanders, general	Outside meanders	Continual
16. W. Elk Cr.	Clay, silt	70–90	1.5	Grasses, shrubs, few trees	General	Where hoof damage, steep banks	Continual
17. Beaver Cr.	Silt	90	2.4	Grasses, few bushes	General	Both banks—overheightened	Continual
18. Brush Creek	Silt	80	1.2–9.1 (top)	Sparse trees	General	Minor, both banks	Continual
19. Unnamed	Clay, silt, sand	40–60	1.2–4.6	Sparse trees, grass, shrubs	Outside meander bends	RB	Frequent
20. Little Skin Cr.	Clay, silt	60	1.5	Grass, shrubs, trees (moderately dense)	Outside meander bend, general	Mostly d/s	Occasional

LB = left bank, RB = right bank, u/s = upstream, d/s = downstream, R. = River, Cr. = Creek

48

Table 7. River bank data, continued.

River	Material	Bank Angle (deg)	Bank Height (m)	Vegetation	Erosion Location	Bank Failure Locations	Exposed or Bare Banks
21. Unnamed	Clay, silt (main), sand	70–90	0.9	Grass, shrubs, dense trees	General	None	Minor
22. Little Cypress Cr.	Clay, silt	90	0.9–3.7	Grass, dense trees	General	Outside bend	Continuous
23. Unnamed	Clay, silt, sand	80	2.4	Grass, dense trees	General, outside bends	General—overheightened banks	Frequent
24. Unnamed	Clay, silt	60–70	4.6 (levee)	Grass, dense trees	General	None	Occasional
25. Honey Run	Clay, silt	60–90	0.9–1.8 (top of mass wasting)	Grass, dense trees	General	Along RB, especially where vegetation removed	Occasional
26. South Fork	Clay, silt	70–80	2.4	Shrubs, sparse trees	General	Along both banks	Frequent
27. East Fork	Clay, silt	70	1.2	Grass	Outside bend, general	None	Occasional
28. Unnamed	Clay, silt	60–90	0.8	Shrubs, dense trees	General (minor u/s)	None	Occasional
29. McKown Cr.	Clay, silt	60–90	1.2 (u/s), 0.8 (d/s)	Grass u/s, d/s grass on left bank, one row of trees on right	General	U/s only along banks	Occasional
30. Wolf Run	Clay, silt, gravel	60–90	1.2	Grass	General, outside, inside bends	Both banks	Frequent
31. Alligator Cr.	Silt, sand	LB vertical, RB moderate	0.8	LB grass, RB dense trees	LB none (concrete), RB outside meander bend (minor)	None	None

LB = left bank, RB = right bank, u/s = upstream, d/s = downstream, R. = River, Cr. = Creek

49

Table 7. River bank data, continued.

River	Material	Bank Angle (deg)	Bank Height (m)	Vegetation	Erosion Location	Bank Failure Locations	Exposed or Bare Banks
32. Peace R.	Sand	Steep outside bend, moderate elsewhere	3	Dense trees	Opposite bar, outside meander bends	Left bank, outside bends	Minor both banks
33. Blackrock Run	Clay, silt, sand	50–80	0.9–2.7	Shrubs, trees	Outside meander, opposite bar	General— overheightened banks	Occasional
34. Indian Run	Clay, silt	40–80	0.8–1.8	Sparse trees LB, grass RB	General, opposite obstructions	RB where no vegetation	Occasional
35. Middle Patuxent R.	Clay, silt	40–60	1.5–1.8	Shrubs, dense trees in good condition	Outside meander bends, general	Limited on RB where high banks slough	Frequent on both banks
36. Hammond Branch	Clay, silt	60–80	1.5	Grasses, sparse trees, falling into stream, poor condition	Outside meander, general	Minor	Continuous
37. Atherton Tributary	Clay, silt	60–80	2.5–6	Sparse trees LB, dense trees RB, falling trees both banks	Outside meander	None u/s, RB d/s where trees removed	Occasional
38. Stocketts Run	Clay, silt	30–50, 80–90 outside bends	0.8–1.8	Shrubs, dense trees, falling on either side	Outside meander, general	None	Frequent
39. Mill Stream Branch	Clay, silt	40–60	0.9–2.4	Sparse falling trees LB, dense falling trees RB	General	None	Occasional

LB = left bank, RB = right bank, u/s = upstream, d/s = downstream, R. = River, Cr. = Creek

Table 7. River bank data, continued.

River	Material	Bank Angle (deg)	Bank Height (m)	Vegetation	Erosion Location	Bank Failure Locations	Exposed or Bare Banks
40. Kent Co. Tributary	Clay (minor gravel toward bottom)	30–80 (lower where bank failed)	2.4	Grasses, sparse falling trees	General	Both banks, continuous	Frequent
41. Morgan Cr.	Clay, silt	30–80, ragged	0.9	Shrubs, sparse trees, trees upright on both banks	General, ragged, irregular banks	Minor	Continuous
42. Little Elk Cr.	Clay, silt, sand	25–34	0.8–0.9	Shrubs, trees, trees upright in good shape	Outside meander bend	None	Occasional
43. Big Beaver Cr.	Clay, silt, sand	Vertical u/s, 30–40 d/s	1.2–2.1 (highly variable)	Grasses u/s, trees beyond 122 m u/s	Outside meander, opposite bar/obstruction	Both banks everywhere for 122 m u/s	Continuous
44. Buffalo Run	Clay, silt	Steep	0.8	Grass, shrubs	General fluvial	RB (ragged)	Frequent on right
45. Roaring Run	Clay, silt, cobbles, boulders	Steep	0.8 (cliff on left)	Rhododendron, trees upright, good shape	Outside meander bend	None	Rare
46. Potter Run	Clay, silt	Moderate to steep	0.6	Trees in good shape on left, grass on right	General fluvial on right	RB along grass	Where grass only
47. Bentley Cr.	Silt, sand, gravel	70–90	0.9	Grass, few trees	General fluvial	Minimal	Frequent
48. N 48	Clay, silt	Moderate to steep	0.6	Very dense trees, healthy, minor grass	General, outside meander bend	None	Minor along reach

LB = left bank, RB = right bank, u/s = upstream, d/s = downstream, R. = River, Cr. = Creek

51

Table 7. River bank data, continued.

River	Material	Bank Angle (deg)	Bank Height (m)	Vegetation	Erosion Location	Bank Failure Locations	Exposed or Bare Banks
49. Reids Run	Clay, silt, cobbles	Moderate to steep	0.9	Moderately dense healthy trees; grasses and shrubs	General	None	Minor along reach
50. Piney Cr.	Clay, silt, sand, gravel	Moderate to steep	1.1	RB grass only; LB healthy trees and shrubs	General	Minor on RB	Minor
51. Little Sandy Cr.	Clay, silt	Moderate	0.8	Grass within 152.5 m u/s of bridge; further u/s healthy dense trees	General	Along grassy areas	Moderate under tree roots and along grass
52. Trout Run	Clay, silt	Steep	0.9	Very dense shrubs, dense healthy trees	General, minor	None	None
53. Pootatuck R.	Clay, silt, sand, gravel, cobbles	Moderate	0.8	Trees, leaning on RB, good shape otherwise	Minor general fluvial	Some mass wasting where trees removed	Occasional
54. Mill R.	Silt, sand, gravel	Steep	0.6	Grasses, annuals, trees. Trees leaning slightly, sparse on LB, dense on RB, good shape	General fluvial	None	Occasional
55. Aspetuck R.	Clay, silt, sand	Moderate	1.1	Annuals (ferns), trees. Moderate density in good shape	General fluvial	None	Occasional
56. W. Br. Saugatuck R.	Clay, silt, sand, gravel	Moderate	0.8	Annuals, shrubs, trees in good health	General fluvial, opposite obstructions (significant)	None	Occasional
57. Mianus R.	Clay, silt, sand, gravel	Moderate to steep	0.9	Annuals, shrubs, trees moderately dense in good shape	General fluvial	None	Minor

LB = left bank, RB = right bank, u/s = upstream, d/s = downstream, R. = River, Cr. = Creek

PHYSIOGRAPHIC REGIONAL OBSERVATIONS

Thirteen physiographic regions and subregions were included in the data set. A wide variety of land uses were observed in the various regions, including natural, agricultural, grazed, rural, and suburban. Width-to-depth ratios also varied widely, from 5 to 24. As expected, the largest ratios were associated with braided or semibraided channels. Very low width-to-depth ratios were associated with incised streams or those that had been engineered. Bed materials varied from very fine materials (silt and very fine sand) in the Midwest and coastal areas to coarser materials mostly associated with higher elevation streams. Bank materials varied widely. For example, the banks of the streams in the Central Plains tend to be made up of fine loess and silt deposits, which erode easily. In the Appalachian Plateau region, in contrast, the bank materials are far more cohesive and tend to be less susceptible to high erosion rates even when bank vegetation was limited.

Although stream types vary widely within any of the physiographic regions, certain common characteristics were observed. A general summary of those observations is given here. The photos of the sites are organized by physiographic region in appendix A. Many, but not all, of them are referred to in the discussion below.

Pacific Coastal

There are two striking features of stream channels within the border and lower Californian subregions of the Pacific Coastal region (see appendix A). The first is the wide diversity in types of channels. Streams range from perennial, cascading channels to arroyos. The second feature common to most streams in these subregions is the frequency of human interference and alteration. The channel bed material in most of the streams (other than first-order streams) was predominantly fine-to-medium sand, while the channel banks were sandy. The average width-to-depth ratio was 12.1. The streams tend to be very high energy; they are typically ephemeral, so they carry water only when there is rainfall. Many of these streams (and arroyos) tend to be naturally unstable, particularly in the lateral direction, and have relatively high width-to-depth ratios. Because of the high degree of channel instability and flash flooding in this region, many, if not most, of the channels in suburban to urban settings were either concrete lined or at least heavily armored with rock. Channels in the outlying areas were unlined.

Intermontane

Observations at bridge-stream intersections were collected in the Basin and Range, Colorado Plateau, and Trans Pecos subregions within the Intermontane physiographic region. Many channels within the Colorado Plateau are bedrock or semialluvial channels in which stability is a function of bedrock erosion. In the Basin and Range, however, where the climate is arid to semiarid over much of the area, the streams are ephemeral with high energy, flashy flows (see appendix A). The energy of these streams combined with the highly erosion-prone sand beds and banks creates unstable channels, particularly at bridges. Due to the high sediment load carried by these streams, the width-to-depth ratios are relatively high, with an average of 25.0. Streams in the Trans Pecos region (see appendix A) tend to have high, steep banks or valley walls, which create valley side failures and subsequent failure material to the stream. The bed materials are sand, and banks are comprised of a mix of noncohesive materials, primarily sand, with minor amounts of cohesive silts and clays. The predominant bank vegetation was desert shrubs, and mass wasting was a common form of erosion. The average width-to-depth ratio was 10.5.

Rocky Mountain System

The southern Rocky Mountains were visited in this region (see appendix A). The channels contain large bed and bank material. The streams tend to be stable, transport streams that are less disturbed by human activities than in other physiographic regions. Stream channel banks are a mix of cohesive silts and clays and noncohesive gravels and larger materials. The average width-to-depth ratio was 16.5.

Interior Plains

Three subregions—the Great Plains, Central Lowlands, and Interior Lowlands—were visited within the Interior Plains. In the Great Plains (see appendix A), vegetation in riparian areas and in the flood plains was thick, lush, and dense, except where cattle were permitted to graze. The channel beds were composed of more than 70 percent sand. Bank material was noncohesive silt, loam, and sand, but the thick vegetation helped to keep banks stable. The average width-to-depth ratio was 17.3. Erosion processes within the stream channels are primarily fluvial; observed channel banks were not sufficiently high to create significant mass wasting. Slow to moderate degradation occurred where cattle grazing was permitted.

The Central Lowlands (see appendix A) had silt and loess banks that eroded easily. Many of the streams observed had been straightened in addition to having extensive hoof and/or grazing damage. The channel beds degrade rapidly since the bed material is predominantly silt with some clay and sand. The silt banks then become overheightened, and mass failures result. High, failing banks were common even where a wider riparian buffer existed, but the rate of failure was slower (for example, see figure 7). The average width-to-depth ratio was much lower than that observed in the Great Plains, with an average of 8.3. This may be due in part to channel modifications, such as straightening.

Streams in the Interior Lowlands (see appendix A) seemed less fragile than those in the Central Lowlands due to larger bed material (sand and gravel) and more cohesive materials in their banks (clay and silt). However, where vegetation had been removed, banks failed even when they were not greatly overheightened (see figure 8). A single row of trees in the riparian areas slowed bank failure dramatically. These streams tend to have a low width-to-depth ratio, with an average of about 10.5, and remain stable even when the surrounding land has been disturbed. This may be due to the existence of rock outcrops in the beds and banks.

Interior Highlands

In the combined Ozark-Ouachita Plateau (see appendix A), bed material was larger, containing some gravel. The bank material contains a significant percentage of cohesive clays. Natural erosion occurred at bends with increased mass wasting at bends where vegetation had been removed. Overheightened banks remained stable when more than one row of trees was in place. The average width-to-depth ratio was 8.5.

Appalachian Highlands

Within the Appalachian Highlands, the Appalachian Plateau, Valley and Ridge, and Piedmont regions were visited. In the Appalachian Plateau (see appendix A), bed material was coarser (mainly very coarse gravel to cobbles), with bank material composed of cohesive clay, silt, and minor sand. Critical bank heights appeared to be about 1.5 to 1.8 m, which result in low width-to-depth ratios. For the sites visited, the average ratio was 11.0. Watersheds are heavily forested where vegetation is undisturbed. Stream channel erosion and destabilization occurs through removal of vegetation and/or channel straightening. Overheightened banks may fail, but heal quickly if vegetation is allowed to re-establish; thus, stability tends to be fair at worst.

Stream channels within the Piedmont region (see appendix A) had cohesive banks that could stand at high angles without failure. Bank vegetation, if undisturbed, was dense and provided bank stability with about one river width of woody vegetation. Banks with angles steeper than about 60° tended to have leaning or fallen trees. The potential for debris jams is high. Occasional bedrock outcropping was noted at all streams that were visited. Bed material was sand and gravel with occasional larger material. The average width-to-depth ratio was 15.3.

The Valley and Ridge region of the Appalachian Highlands is comprised of a series of ridges separated by stream valleys. The streams in this region are often very steep, especially coming down from the ridges (see appendix A). Cascade and step-pool morphologies are common. Thus, bed materials are commonly large, such as cobbles and boulders, and often armor the bed. Banks are cohesive clays and silts with some larger materials mixed in, strongly held together by the lush vegetation found in this area. Disturbance to the banks by removal of vegetation may result in ragged, scalloped banks, but erosion of the banks is typically at a relatively slow rate. The average width-to-depth ratio was 15.0.

Figure 7. Failing banks in the Central Lowlands.

Figure 8. Failing banks in the Interior Lowlands.

Coastal Plain

The Coastal Plain covers a very large area of the Atlantic and Gulf coastal areas (see figure 6). Sites were visited along both the Atlantic and Gulf coasts (see appendix A). In both of these areas, researchers observed moderate rates of bed degradation and bank failure. A buffer of at least one river width appeared to be sufficient in most locations to keep banks stable. Where undisturbed, lush vegetation on the banks held the banks in place, resulting in excellent stability, even when banks were nearly vertical. Bed material is typically sand with minor amounts of small gravel, and banks are cohesive with clay, silt, and minor amounts of sand. Because of the cohesive banks, strong vegetative resistance, and degradation, width-to-depth ratios tended to be rather low. Streams in this region are often sluggish due to low slopes and backwater from the bays or estuaries into which they flow. Where banks or the flood plain are disturbed, debris jams are frequent. The average width-to-depth ratios were much lower in the Gulf area (9.0) than in the Atlantic area (13.5).

New England

All of the streams visited in the New England region were located in Connecticut. At all streams, the banks were heavily vegetated with large woody vegetation, providing tremendous stability to the streambanks. The bank materials typically were comprised of some cohesive materials combined with silt, sand and, in some places, gravel and larger particles. The bed materials in the New England region are considerably larger than in the Atlantic Coastal Plain to the south. The sand, gravel, and cobble beds were often armored; the width-to-depth ratios reflected this armored condition with an average value of 24. The channels were all meandering, but with beds transitional between plane beds and pool-riffle beds.

General Observations of Streams at Bridges

Channel stability is a function of levels of disturbance to the water and sediment discharges, and susceptibility of the channels to change. In every physiographic region, the disturbance that caused the greatest damage to the streams was the combination of cattle activity, vegetation removal, and channel straightening. The combined impact of these activities was worst where cattle had direct access to streams. Also, susceptibility of the channel banks to erosion significantly impacted the level of damage. Figures 9 and 10 provide examples of this combination of disturbances. All vegetation has been removed either through farming practices or by cattle grazing. The channel apparently had been straightened to provide better drainage and to maximize land for farming. Not only are cattle grazing in this area, but also they have direct access to the stream. Hoof damage is extensive. The combined disturbances have resulted in stream channel destabilization; the channel bed elevation has degraded and the banks have become overheightened and steepened. Figure 10 shows the eroding channel beneath the single-span bridge.

In many cases, maintaining a riparian buffer of an appropriate width is all that is needed to preserve channel stability. As discussed in the descriptions of the streams channels across the physiographic regions, some regions require only a single row of trees to help maintain stability, while others require a much greater width. This is due to bank materials and the susceptibility of the banks to failure. In the cases where channels are degrading because of channel straightening, cattle grazing, and urbanization effects, a vegetation buffer may not be enough to maintain stability. When the channel degrades, banks can become overheightened and fail through mass wasting. In this case, vegetation may help to slow the rate of failure, but usually cannot prevent collapse of the banks.

Figure 9. Stream impacts due to disturbances, including hoof damage,
vegetation removal, and channel straightening.

Figure 10. Impacts of disturbances at bridge (from figure 9).

Another observation that was frequently made at sites in all physiographic regions was that there was often a distinct change in channel stability upstream and downstream of the bridges. This was caused in every case by a change in property management, as it is common for a road (and, thus, a bridge) to divide property ownership. As an example, unnamed stream N 28 is wooded upstream, with a healthy wide band of upright trees keeping the banks stable (see figure 11). Immediately downstream of the bridge, all trees and other vegetation have been removed, resulting in destabilization of the banks (see figure 12).

Aerial photos were examined for each of the sites using http://terraserver-usa.com (these photos are not included in the report because they are readily available online). The photos were examined to check a larger view of the river, specifically looking at land use in the watershed and flood plain, construction areas, the extent of the riparian buffer, channel straightening, and channel pattern. In most cases, the aerial photos reinforced observations that were made on the ground. In a number of cases, the photos helped put the bridge reach into the perspective of the meander pattern, particularly where the bridge was located between meanders or just downstream of a tight meander. Old abandoned meanders also could be detected sometimes, giving an indication of previous lateral movement. Changes in channel pattern, for example from meandering to braided, can be detected on aerial photos. Examining the photos before or after visiting a site helped provide a rating, especially for the watershed condition factor.

Figure 11. Wooded land upstream of bridge.

Figure 12. Downstream of figure 11, vegetation removed.

EFFECT OF CHANNEL INSTABILITY ON BRIDGES

Unstable channels can cause a variety of problems at bridges; however, this is not necessarily the case. For example, the Mojave River in California (see figure 13) can be considered to be a naturally unstable channel, primarily in the lateral direction, in that there is considerable lateral movement of the channel. The channel bed and banks are noncohesive fine sand that adjust readily to sudden changes in hydrology from a dry condition to flash flooding. However, the bridge at the site that was visited spans a wide section of the flood plain, thus providing room for some lateral migration. In many other sites visited, lateral migration of meanders was a potential threat to bridge abutments. In figure 14, lateral migration of a gentle meander bend has forced the channel against the left abutment. This has, in turn, caused additional local scour at the abutment and undermining of the abutments, and could result in an unstable bridge foundation. Lateral and downstream migration of this meander would have a significant impact on the left abutment.

Figure 13. Mojave River, CA.

Figure 14. Meander migration affecting right abutment,
Hammond Branch, MD.

One of the biggest problems created by channel instability at bridges exists at single-span bridges that are only as wide as the channel. This allows for no or limited lateral or vertical adjustments of the channel. As an example, figure 15 shows a single-span bridge across a channel that is both degrading and widening. Significant widening will result in undermining of the abutment walls.

Even for channels that are unstable, the bridge may not be in danger if adequate structural redundancy is in place. Thus, an observation of channel instability is not a sufficient condition for impending structural failure. The bridge inspector must consider what impact, if any, a channel that is deemed unstable will have during the time period between inspections, especially in the event of a large hydrologic event.

Channel stabilization measures at bridges are quite common. Given the small right-of-way at most bridges, the measures typically are placed directly at the bridge and perhaps a short distance upstream or downstream. By far, the most common type of stabilization measure observed at these sites was riprap. In some cases, the riprap appeared to be effective in holding the bank in place at the bridge. In other cases, however, riprap did not appear to be effective without significant maintenance. For example, at S.R. 445 over Roaring Run in Pennsylvania, there is a high riprap wall composed of graded riprap with a median size of about 152–229 mm (see figure 16). The purpose of the wall is to prevent lateral migration of the tight meander bend just upstream of the bridge. The wall has a bank angle of about 70°. This configuration of loose, undersized riprap in such a steep arrangement has little chance of withstanding the high shear stresses imposed on it at high flows as the high gradient stream makes this tight bend. There is already evidence of riprap wall failure, as much of the stone is deposited in the stream channel just upstream of the bridge. In other cases, stabilization efforts seem to work quite well. As an example, a cross vane has been installed just downstream of the S.R. 144 bridge over Potter Run in Pennsylvania (see figure 17). The cross vane causes the flow to pool just upstream and under the bridge, slowing the high velocity and minimizing scour under the bridge and along the banks.

RELATIONSHIP BETWEEN CHANNEL STABILITY AND SCOUR AT BRIDGES

In HEC-18, scour is defined as having three vertical components: local, contraction, and bed degradation. Local and contraction scours are caused by the bridge and occur within close vicinity of the bridge. Bed degradation, on the other hand, is not caused by the bridge and may be reach-wide or even systemwide. Channel instability includes bed degradation, but also comprises other components, based on the definition given previously, such as channel widening, lateral migration, and bed aggradation. At bridges, channel instabilities can cause:

- Channel bed degradation, which may undermine the bridge foundations.
- Channel widening, which can undermine and outflank bridge abutments and piles in the flood plain.
- Lateral migration, which can undermine abutments and permit local scour to be far more productive as the channel thalweg nears an abutment.

Channel aggradation in itself is not usually detrimental to the bridge structure, but it can lead to increased flooding and channel widening. At many of the bridges observed during this project, narrow, single-span bridges often were impacted more because small lateral movements of the channel could press the stream thalweg up against one abutment, increasing the local scour at that abutment.

Figure 15. Single-span bridge over unstable channel.

Figure 16. Riprap stabilization wall along Roaring Run, PA.

Figure 17. Cross vane downstream of bridge over Potter Run, PA.

4. ASSESSING CHANNEL STABILITY

Based on the studies described in a previous section as well as on the observations made at bridges across numerous physiographic regions, a group of parameters that indicate channel stability can be selected. First, however, it is necessary to redefine stream channel stability in light of bridge engineering issues. For this purpose, a stable channel is defined as follows, based on Knox and modified for use at bridges:[15]

> *A stable channel in the vicinity of a bridge is one in which the relationship between geomorphic process and form is stationary and the morphology of the system remains relatively constant over the short-term (one to two years), over a short distance upstream and downstream from bridge, and with minimal lateral movement.*

Although lateral migration of a stream channel can be considered normal and stable within a geomorphic definition of channel stability, it is detrimental to bridge safety and is, therefore, considered in the stability definition used here. The distance upstream and downstream of the bridge that should be considered in a stability assessment depends on the problem, the channel, and the bridge. However, it is acknowledged that a bridge inspector will not typically walk more than a few hundred feet in either direction. That stated, it should be noted that without walking well upstream and downstream of the bridge, channel instabilities, such as knickpoints, that are migrating toward the bridge area may be overlooked. Remember that the objective here is only to assess stream stability in the short term, as inspections of bridges over water are required every 2 years. Thus, it is not necessary to develop a complex method to examine the history or future of channel adjustments over a long time period. It is necessary, however, for each inspector to review previous stability analyses at the bridge of interest to determine whether any unstable trends are developing.

A stability assessment program for bridge inspections should be: (1) brief so that it can be completed rapidly; (2) simple in that extensive training is not required (although some training will be required); (3) based on sound indicators as discussed in the literature review; and (4) based on the needs of the bridge engineering community.

One way to insure that all aspects of channel stability are included is to start at the watershed or regional level and focus in on vertical and lateral aspects of the channel, following the concepts of Thorne et al.[7] and Montgomery and MacDonald.[20] Thus, at the broader level, watershed and flood plain activities as well as characteristics, flow habit, channel pattern and type, and entrenchment are selected as appropriate indicators. At the channel level, indicators such as bed material consolidation and armoring, bar development, and obstructions are used. Indicators of bank stability include bank material, angle, bank and riparian vegetation, bank (fluvial) cutting, and mass wasting (geotechnical failure). Finally, the position of the bridge relative to the channel can be indicated by meander impact point and alignment. In the previous method, the ratio of the average boundary shear stress to the critical shear stress for sediment movement had been found to be important; however, average shear conditions do not necessarily indicate processes that are occurring.[1] Also, critical shear stress is not a reliable number. In addition, it is difficult to measure and quantify as part of a rapid assessment. Therefore, the shear stress ratio is not used as a stability indicator in this current assessment method. In its place, bed material and percent of sand are used. These results are based on the Wilcock and Kenworthy study of bed material movement as a function of sand fractions.[86]

The 13 indicators identified for this study are listed in table 8. For each indicator, a rating of poor, fair, good, or excellent can be assigned based on descriptors listed in the table. After a rating is assigned for each of the indicators, an overall rank is obtained by summing the 13 ratings. Several assumptions are implicit in this method of obtaining an overall rank. First, all indicators are weighted equally. This assumption was tested by assigning weights to each of the indicators and creating a weighted score for every bridge where observations were made. The results showed that the weighted indicators yielded the same results as the equally weighted indicators. Thus, there was no advantage in using weights. Second, this method implies that each indicator is independent of all others. While it is possible that some correlation exists between several of the indicators, an attempt was made to select indicators that independently describe various aspects of channel stability; thus, correlation effects were judged to be insignificant. Third, the summing of the ratings implies a linear scheme. The impact of this is not precisely known; however, given that weighted ratings provided no change in the overall results, it can be assumed that the linearity will also not affect the results significantly.

Table 9 provides the rating results for each of the 13 stability indicators at all of the bridges where observations were made. Using the same 13 indicators for streams in all physiographic regions adequately described channel conditions at each of these sites. The sums of the ratings also are given in table 9. These overall rankings were then rated as excellent, good, fair, or poor. The division of the overall rankings among a single set of category divisions provided limited sensitivity to streams in some stream channel classifications and physiographic regions. Thus, it was desirable to rank the stability based on stream type. Given that the Montgomery-Buffington classification method is based on processes as well as physical characteristics, this scheme was used to provide additional sensitivity to the method. Since cascade and step-pool streams are both transport streams and are not sensitive to changes in sediment or water discharge, these streams were given a separate category of rankings. Plane-bed, pool-riffle, and dune-ripple streams, along with engineered channels, were given a second category as primarily response-type streams. Finally, braided streams were placed in a third category, as these represent a type of stream that is very sensitive to changes in sediment and water discharge and are primarily located in the western and southwestern regions of the United States. These divisions also agree loosely with the stability assessment method that Rosgen developed. (Rosgen has divisions according to stream type, resulting in 42 divisions. This implies a level of sensitivity for which there is no explanation given. It also provides an unwieldy and cumbersome accounting of rankings and tables.) Tables 10–12 provide the range of rankings for excellent, good, fair, and poor ratings of stability for each of the three divisions of stream channels. The final rankings, in terms of excellent, good, fair, and poor, are given in table 9.

Table 8. Stability indicators, descriptions, and ratings.*

Stability Indicator	Ratings			
	Excellent (1–3)	Good (4–6)	Fair (7–9)	Poor (10–12)
1. Watershed and flood plain activity and characteristics	Stable, forested, undisturbed watershed	Occasional minor disturbances in the watershed, including cattle activity (grazing and/or access to stream), construction, logging, or other minor deforestation. Limited agricultural activities	Frequent disturbances in the watershed, including cattle activity, landslides, channel sand or gravel mining, logging, farming, or construction of buildings, roads, or other infrastructure. Urbanization over significant portion of watershed	Continual disturbances in the watershed. Significant cattle activity, landslides, channel sand or gravel mining, logging, farming, or construction of buildings, roads, or other infrastructure. Highly urbanized or rapidly urbanizing watershed
2. Flow habit	Perennial stream with no flashy behavior	Perennial stream or ephemeral first-order stream with slightly increased rate of flooding	Perennial or intermittent stream with flashy behavior	Extremely flashy; flash floods prevalent mode of discharge; ephemeral stream other than first-order stream
3. Channel pattern	Straight to meandering with low radius of curvature; primarily suspended load	Meandering, moderate radius of curvature; mix of suspended and bed loads; well-maintained engineered channel	Meandering with some braiding; tortuous meandering; primarily bed load; poorly maintained engineered channel	Braided; primarily bed load; engineered channel that is not maintained
4. Entrenchment/ channel confinement	Active flood plain exists at top of banks; no sign of undercutting infrastructure; no levees	Active flood plain abandoned, but is currently rebuilding; minimal channel confinement; infrastructure not exposed; levees are low and set well back from the river	Moderate confinement in valley or channel walls; some exposure of infrastructure; terraces exist; flood plain abandoned; levees are moderate in size and have minimal setback from the river	Knickpoints visible downstream; exposed water lines or other infrastructure; channel-width-to-top-of-banks ratio small; deeply confined; no active flood plain; levees are high and along the channel edge

*Range of values in ratings columns provide possible rating values for each factor

H = horizontal, V = vertical, Fs = fraction of sand, S = slope, w/y = width-to-depth ratio

Table 8. Stability indicators, descriptions, and ratings, continued.

Stability Indicator	Ratings			
	Excellent (1–3)	Good (4–6)	Fair (7–9)	Poor (10–12)
5. Bed material Fs = approximate portion of sand in the bed	Assorted sizes tightly packed, overlapping, and possibly imbricated. Most material > 4 mm. Fs < 20%	Moderately packed with some overlapping. Very small amounts of material < 4 mm. 20 < Fs < 50%	Loose assortment with no apparent overlap. Small to medium amounts of material < 4 mm. 50 < Fs < 70%	Very loose assortment with no packing. Large amounts of material < 4 mm. Fs > 70%
6. Bar development	For S < 0.02 and w/y > 12, bars are mature, narrow relative to stream width at low flow, well vegetated, and composed of coarse gravel to cobbles. For S > 0.02 and w/y < 12, no bars are evident	For S < 0.02 and w/y > 12, bars may have vegetation and/or be composed of coarse gravel to cobbles, but minimal recent growth of bar evident by lack of vegetation on portions of the bar. For S > 0.02 and w/y < 12, no bars are evident	For S < 0.02 and w/y > 12, bar widths tend to be wide and composed of newly deposited coarse sand to small cobbles and/or may be sparsely vegetated. Bars forming for S > 0.02 and w/y < 12	Bar widths are generally greater than 1/2 the stream width at low flow. Bars are composed of extensive deposits of fine particles up to coarse gravel with little to no vegetation. No bars for S < 0.02 and w/y > 12
7. Obstructions, including bedrock outcrops, armor layer, LWD jams, grade control, bridge bed paving, revetments, dikes or vanes, riprap	Rare or not present	Occasional, causing cross currents and minor bank and bottom erosion	Moderately frequent and occasionally unstable obstructions, cause noticeable erosion of the channel. Considerable sediment accumulation behind obstructions	Frequent and often unstable, causing a continual shift of sediment and flow. Traps are easily filled, causing channel to migrate and/or widen
8. Bank soil texture and coherence	Clay and silty clay; cohesive material	Clay loam to sandy clay loam; minor amounts of noncohesive or unconsolidated mixtures; layers may exist, but are cohesive materials	Sandy clay to sandy loam; unconsolidated mixtures of glacial or other materials; small layers and lenses of noncohesive or unconsolidated mixtures	Loamy sand to sand; noncohesive material; unconsolidated mixtures of glacial or other materials; layers or lenses that include noncohesive sands and gravels

H = horizontal, V = vertical, Fs = fraction of sand, S = slope, w/y = width-to-depth ratio

Table 8. Stability indicators, descriptions, and ratings, continued.

Stability Indicator	Ratings			
	Excellent (1–3)	Good (4–6)	Fair (7–9)	Poor (10–12)
9. Average bank slope angle (where 90° is a vertical bank)	Bank slopes < 3H:1V (18°) for noncohesive or unconsolidated materials to < 1:1 (45°) in clays on both sides	Bank slopes up to 2H:1V (27°) in noncohesive or unconsolidated materials to 0.8:1 (50°) in clays on one or occasionally both banks	Bank slopes to 1H:1V (45°) in noncohesive or unconsolidated materials to 0.6:1 (60°) in clays common on one or both banks	Bank slopes over 45° in noncohesive or unconsolidated materials or over (60°) in clays common on one or both banks
10. Vegetative or engineered bank protection	Wide band of woody vegetation with at least 90% density and cover. Primarily hard wood, leafy, deciduous trees with mature, healthy, and diverse vegetation located on the bank. Woody vegetation oriented vertically. In absence of vegetation, both banks are lined or heavily armored	Medium band of woody vegetation with 70–90% plant density and cover. A majority of hard wood, leafy, deciduous trees with maturing, diverse vegetation located on the blank. Woody vegetation oriented 80–90° from horizontal with minimal root exposure. Partial lining or armoring of one or both banks	Small band of woody vegetation with 50–70% plant density and cover. A majority of soft wood, piney, coniferous trees with young or old vegetation lacking in diversity located on or near the top of bank. Woody vegetation oriented at 70–80° from horizontal, often with evident root exposure. No lining of banks, but some armoring may be in place on one bank	Woody vegetation band may vary depending on age and health with less than 50% plant density and cover. Primarily soft wood, piney, coniferous trees with very young, old and dying, and/or monostand vegetation located off of the bank. Woody vegetation oriented at less than 70° from horizontal with extensive root exposure. No lining or armoring of banks
11. Bank cutting	Little or none evident. Infrequent raw banks, insignificant percentage of total bank	Some intermittently along channel bends and at prominent constrictions. Raw banks comprise minor portion of bank in vertical direction	Significant and frequent on both banks. Raw banks comprise large portion of bank in vertical direction. Root mat overhangs	Almost continuous cuts on both banks, some extending over most of the banks. Undercutting and sod-root overhangs

H = horizontal, V = vertical, Fs = fraction of sand, S = slope, w/y = width-to-depth ratio

67

Table 8. Stability indicators, descriptions, and ratings, continued.

Stability Indicator	Ratings			
	Excellent (1–3)	Good (4–6)	Fair (7–9)	Poor (10–12)
12. Mass wasting or bank failure	No or little evidence of potential or very small amounts of mass wasting. Uniform channel width over the entire reach	Evidence of infrequent and/or minor mass wasting. Mostly healed over with vegetation. Relatively constant channel width and minimal scalloping of banks	Evidence of frequent and/or significant occurrences of mass wasting that can be aggravated by higher flows, which may cause undercutting and mass wasting of unstable banks. Channel width quite irregular, and scalloping of banks is evident	Frequent and extensive mass wasting. The potential for bank failure, as evidenced by tension cracks, massive undercuttings, and bank slumping, is considerable. Channel width is highly irregular, and banks are scalloped
13. Upstream distance to bridge from meander impact point and alignment	More than 35 m; bridge is well-aligned with river flow	20–35 m; bridge is aligned with flow	10–20 m; bridge is skewed to flow, or flow alignment is otherwise not centered beneath bridge	Less than 10 m; bridge is poorly aligned with flow

H = horizontal, V = vertical, Fs = fraction of sand, S = slope, w/y = width-to-depth ratio

68

Table 9. Stability assessment ratings for each factor.

| Stream | Indicator | | | | | | | | | | | | | Total | Rating Based on Tables 11–13 |
	1	2	3	4	5	6	7	8	9	10	11	12	13		
Saline R.	6	9	7	3	9	6	4	9	3	4	2	2	3	67	Good
SF Solomon R.	6	10	8	4	11	5	2	12	7	8	2	3	3	81	Good
N. Rush Cr.	5	9	6	7	9	2	2	9	6	10	5	5	9	84	Good
Arkansas R.	4	3	5	2	5	2	5	9	2	6	3	4	4	54	Good
Tomichi Cr.	7	3	5	2	9	3	4	7	5	5	3	2	7	62	Good
Murietta Cr.	12	12	9	7	11	5	2	11	9	10	8	7	3	106	Fair
Jacalitos Cr.	9	12	10	7	11	8	3	11	8	10	6	7	8	110	Fair
Dry Cr.	3	7	3	7	3	2	9	3	8	5	4	4	5	63	Good
Dutch Bill Cr.	2	8	5	7	7	10	5	8	8	3	6	4	10	83	Good
Buena Vista Cr.	8	11	10	7	12	10	2	11	10	11	10	10	4	116	Fair
Mojave R.	10	12	12	6	12	12	5	12	7	12	12	11	8	131	Poor
Rt. 66 Wash	10	12	10	6	8	10	9	11	10	12	12	11	11	132	Poor
Sacramento Wash	9	12	10	6	10	10	9	11	12	10	12	9	4	124	Fair
Rio San Jose	8	7	4	9	9	4	6	10	10	9	9	9	6	100	Fair
Rio Puerco	8	7	6	10	11	10	9	12	12	11	12	12	10	130	Poor
W. Elk Creek	7	4	5	8	10	6	9	7	11	8	9	10	3	97	Fair
Beaver Cr.	12	4	6	12	11	10	10	12	12	11	12	12	3	127	Poor
Brush Cr.	10	6	6	10	8	4	7	11	10	9	7	8	3	99	Fair
N 19	7	7	6	8	10	10	7	7	8	7	8	7	8	100	Fair

R. = River, Cr. = Creek

Table 9. Stability assessment ratings for each factor, continued.

Stream	Indicator													Total	Rating Based on Tables 11–13
	1	2	3	4	5	6	7	8	9	10	11	12	13		
Little Skin Cr.	6	5	6	5	5	6	9	7	9	9	6	10	8	91	Fair
N 21	6	7	3	6	9	4	8	8	10	7	5	3	10	86	Fair
Little Cypress Cr.	8	6	6	11	10	8	7	6	12	7	12	9	8	110	Fair
N 23	5	4	7	12	10	11	9	6	12	9	7	12	8	112	Fair
N 24	7	5	6	10	11	6	6	5	10	9	5	6	3	89	Fair
Honey Run	4	3	2	6	5	4	4	5	10	9	5	11	10	78	Good
South Fork	3	3	3	7	7	2	4	4	8	8	6	6	3	64	Good
East Fork	8	5	6	7	6	4	5	5	10	11	6	5	10	88	Fair
N 28	9	5	6	7	3	4	8	4	10	8	7	4	4	79	Good
McKnown Cr.	3	3	5	7	3	3	7	3	9	11	7	6	4	71	Good
Wolf Run	3	3	6	9	2	4	9	9	11	11	11	11	9	98	Fair
Alligator Cr.	8	2	4	5	3	5	5	8	5	2	2	2	6	57	Good
Peace R.	4	2	8	2	12	5	8	11	5	5	4	7	4	77	Good
Blackrock Run	7	4	5	8	7	7	5	7	10	8	4	7	5	84	Good
Indian Run	5	3	8	7	5	8	8	5	8	10	4	9	10	90	Fair
Middle Patuxent R.	5	2	5	4	3	3	4	5	9	5	8	7	7	67	Good
Hammond Branch	11	9	8	8	9	9	7	6	10	9	10	9	11	116	Fair
Atherton Tributary	5	6	5	6	6	5	4	5	8	8	6	6	4	74	Good
Stocketts Run	3	3	4	6	7	8	5	2	7	5	8	4	11	73	Good

R. = River, Cr. = Creek

70

Table 9. Stability assessment ratings for each factor, continued.

Stream	Indicator													Total	Rating Based on Tables 11–13
	1	2	3	4	5	6	7	8	9	10	11	12	13		
Mill Stream Branch	6	4	3	7	10	8	5	3	8	7	4	4	3	72	Good
Kent County Tributary	4	3	4	10	10	9	7	3	11	9	9	10	9	98	Fair
Morgan Cr.	8	5	5	4	9	5	4	3	9	9	10	6	11	88	Fair
Little Elk Cr.	5	2	3	2	2	3	3	4	5	3	4	2	8	46	Excellent
Big Beaver Cr.	11	5	5	4	10	12	6	5	10	12	12	10	10	112	Fair
Buffalo Run	7	4	6	5	5	4	5	2	9	11	9	7	4	78	Good
Roaring Run	2	2	2	2	2	5	3	3	4	3	3	1	8	40	Excellent
Potter Run	3	3	3	3	1	2	4	2	6	9	8	4	4	52	Good
Bentley Cr.	10	9	10	7	9	6	5	12	12	8	11	8	12	119	Fair
N 48	2	3	3	5	1	2	4	1	5	1	4	2	8	41	Excellent
Reids Run	3	3	3	5	1	4	4	3	6	5	4	2	7	50	Good
Piney Cr.	4	3	4	4	3	5	4	5	7	9	4	2	9	63	Good
L. Sandy Cr.	6	4	4	6	3	5	3	3	8	10	6	8	8	74	Good
Trout Run	4	3	3	3	4	3	3	3	6	1	2	1	7	43	Excellent
Pootatuck R.	4	3	3	4	1	2	4	3	5	5	5	4	8	51	Good
Mill R.	3	2	5	3	6	2	4	8	6	6	6	2	12	65	Good
Aspetuck R.	5	3	3	5	7	4	4	7	5	5	6	2	3	59	Good
W. Br. Saugatuck R.	6	3	7	3	5	2	11	6	3	3	6	1	8	64	Good
Mianus R.	3	2	3	3	4	3	3	4	5	4	4	1	5	44	Excellent

R. = River, Cr. = Creek

71

Table 10. Overall rankings for pool-riffle, plane-bed, dune-ripple, and engineered channels.

Category	Ranking, R
Excellent	$R < 49$
Good	$49 \leq R < 85$
Fair	$85 \leq R < 120$
Poor	$120 \leq R$

Table 11. Overall rankings for cascade and step-pool channels.

Category	Ranking, R
Excellent	$R < 41$
Good	$41 \leq R < 70$
Fair	$70 \leq R < 98$
Poor	$98 \leq R$

Table 12. Overall rankings for braided channels.

Category	Ranking, R
Excellent	N/A
Good	$R < 94$
Fair	$94 \leq R < 129$
Poor	$129 \leq R$

Table 13. Vertical versus lateral stability.

Stream	Lateral	Vertical	Lateral Fraction	Vertical Fraction
Saline R.	23	18	0.32	0.50
SF Solomon R.	35	20	0.49	0.56
N. Rush Cr.	44	18	0.61	0.50
Arkansas R.	28	9	0.39	0.25
Tomichi Cr.	29	14	0.40	0.39
Murietta Cr.	48	23	0.67	0.64
Jacalitos Cr.	50	26	0.69	0.72
Dry Cr.	29	12	0.40	0.33
Dutch Bill Cr.	39	24	0.54	0.67
Buena Vista Cr.	56	29	0.78	0.81
Mojave R.	62	30	0.86	0.83
Rt. 66 Wash	67	24	0.93	0.67
Sacramento Wash	58	26	0.81	0.72
Rio San Jose	53	22	0.74	0.61
Rio Puerco	69	31	0.96	0.86
W. Elk Creek	48	24	0.67	0.67
Beaver Cr.	62	33	0.86	0.92
Brush Cr.	48	22	0.67	0.61
N 19	45	28	0.63	0.78
Little Skin Cr.	49	16	0.68	0.44
N 21	43	19	0.60	0.53
Little Cypress Cr.	54	29	0.75	0.81
N 23	54	33	0.75	0.92
N 24	38	27	0.53	0.75
Honey Run	50	15	0.69	0.42
South Fork	35	16	0.49	0.44
East Fork	47	17	0.65	0.47

R. = River, Cr. = Creek

Table 13. Vertical versus lateral stability, continued.

Stream	Lateral	Vertical	Lateral Fraction	Vertical Fraction
N 28	37	14	0.51	0.39
McKnown Cr.	40	13	0.56	0.36
Wolf Run	62	15	0.86	0.42
Alligator Cr.	25	13	0.35	0.36
Peace R.	36	19	0.50	0.53
Blackrock Run	41	22	0.57	0.61
Indian Run	46	20	0.64	0.56
Middle Patuxent R.	41	10	0.57	0.28
Hammond Branch	55	26	0.76	0.72
Atherton Tributary	37	17	0.51	0.47
Stocketts Run	37	21	0.51	0.58
Mill Stream Branch	29	25	0.40	0.69
Kent Co. Tributary	51	29	0.71	0.81
Morgan Creek	48	18	0.67	0.50
Little Elk Cr.	26	7	0.36	0.19
Big Beaver Cr.	59	26	0.82	0.72
Buffalo Run	42	14	0.58	0.39
Roaring Run	22	9	0.31	0.25
Potter Run	33	6	0.46	0.17
Bentley Cr.	63	22	0.88	0.61
N 48	21	8	0.29	0.22
Reids Run	27	10	0.38	0.28
Piney Cr.	36	12	0.50	0.33
L. Sandy Cr.	43	14	0.60	0.39
Trout Run	20	10	0.28	0.28
Pootatuck R.	30	7	0.42	0.19
Mill R.	40	11	0.56	0.31

R. = River, Cr. = Creek

Table 13. Vertical versus lateral stability, continued.

Stream	Lateral	Vertical	Lateral Fraction	Vertical Fraction
Aspetuck R.	28	16	0.39	0.44
W. Br. Saugatuck R.	27	10	0.38	0.28
Mianus R.	23	10	0.32	0.28

R. = River, Cr. = Creek

HEC-20 suggests that the lateral and vertical stability be examined as well as the overall stability. The indicators in table 8 can be divided into those that indicate vertical stability and those that indicate lateral stability. Results are given in table 13 in which vertical stability is described by indicators 4–6, while lateral stability is described by indicators 8–13. Each of the lateral and vertical stability ratings were normalized by the total number of points possible in each category so that they could be represented as a fraction and more readily compared. Thus, the lateral score was divided by 72 and the vertical score by 36. If the lateral score fraction is greater than the vertical score fraction, then it can be expected that the channel instability is primarily in the lateral direction. As an example, the Route 66 Wash is rated as "poor." However, the lateral score fraction is significantly higher than the vertical score fraction (0.93 versus 0.67), indicating that lateral instability is dominant. If, on the other hand, the vertical score fraction is greater than the lateral, then bed degradation is the dominant source of instability. An example of this type of scores is given by Wolf Run, for which the vertical score fraction is about double the lateral score fraction, indicating primarily vertical instability. If both scores are high, then the channel is unstable due to both lateral and vertical processes. For example, Beaver Creek has lateral and vertical fractions of 0.86 and 0.92, respectively. This indicates that the channel is both degrading and widening. The processes may be ongoing simultaneously or they may be occurring differentially. This is frequently the case—as a knickpoint moves upstream, the channel banks respond by collapsing and widening, then another knickpoint moves through, and the process repeats. If both scores are low, this indicates minimal instability in either direction. For example, Alligator Creek has similar scores in both lateral and vertical categories, indicating healthy adjustments in both directions.

Occasionally, rating each of the 13 factors for a particular bridge will result in one factor which stands out as being much higher (worse) than the others. For example, Little Elk Creek received an excellent as the overall rating. All of the assessment factors received scores between 2 and 5, except for the alignment factor (#13). This factor was given a rating of 8 due to the fact that the right abutment of the bridge was located just downstream of the outside of a gentle meander bend. The meander bend appears to be migrating at a very slow rate; this is based on observations that there is undercutting of tree roots on the right bank, but all trees are oriented vertically. Although the rate of lateral migration appears to be slow, it is worth noting and making additional observations during future inspections.

In collecting the data and observations for this method, the engineer or other inspector should walk some distance upstream and downstream from the bridge, rather than just observe from the bridge itself. The appropriate distance, however, depends on several factors, such as uniformity of stream conditions, magnitude of disturbances along the banks, in the flood plain, or in the watershed, time available, and accessibility. Ideally, the observer should walk at least 10 channel widths upstream and downstream of the bridge. Although it is possible to establish stability conditions in less distance, the more of the stream that is observed, the better understanding the observer will have of causes, processes, and rates of change.

Bridges often divide property and sometimes divide geomorphic features or regions. Thus, conditions upstream and downstream of the bridge may be significantly different. In this case, it may be necessary to conduct separate analyses upstream and downstream. Unless the disturbance downstream of the bridge is traveling upstream, as in knickpoint migration or lateral migration of an adjacent meander, then the conditions downstream will be unlikely to affect the bridge, and more emphasis should be placed on the upstream conditions.

5. MODIFICATIONS OF THORNE'S RECONNAISSANCE SHEETS

The stream stability assessment method developed in chapter 4 is self-contained and does not require any other data or formal method of data collection other than the descriptors given in table 8. However, it is prudent to develop field forms that help observers focus attention on specific aspects of a stream, be consistent in those observations, and systematically record their observations. For this purpose, the Thorne reconnaissance field sheets are the best method available for systematically collecting stream channel data. There are several problems with using the reconnaissance field sheets developed by Thorne for highway-related purposes.[2] First, it is not clear that this highly detailed and time-consuming level of data collection is necessary to assess stream stability or stream response at a bridge. In addition, bridge inspectors generally cannot and will not take the time to collect this level of data, as it is out of line with the overall inspection process. Second, even when all of the detailed data are collected, there is no guidance within the method for interpreting the data. Third, few inspectors and hydraulic engineers have appropriate backgrounds to identify geological parameters, such as rock type. To develop a rapid stability assessment method, a simplified version of the Thorne reconnaissance sheets should be created specifically for use in training and data collection associated with assessing stream stability at bridges.

The stream stability method developed for this report is based on data collected through the reconnaissance. However, given that bridge inspection requires only assessment of stream stability in the short term and since each bridge is inspected at least every 2 years, data collection requirements can be simplified to reflect this less detailed and rapid assessment. In addition, several aspects of the sheets required minimal revision. Interpretive observations, while critical to communicating between observers, are neglected in the revised sheets because, in a rapid assessment, most of the qualitative data are collected by an inspector who is interpreting what he or she observes. All inspections require written reports in which the inspector provides his or her judgment on the status of the overall bridge condition and maintenance needs. In addition, items on the sheet that cannot be assessed in a very brief site visit are excluded from the revised sheets. The simplified and revised data collection sheets, based strongly on Thorne's reconnaissance sheets, are given in figures 18–20.

In addition to the stability assessment, keeping a record of channel dimensions upstream and downstream of the bridge will provide a history of changes in width and depth. Although detailed surveys are not part of a bridge inspection, a simple measurement of station and elevation upstream of the bridge taken annually will provide adequate cross-sectional information to assess longer term changes. Without this information, gradual but continual changes in the channel may be overlooked.

```
┌─────────────────────────────────────────────────────────────────────────┐
│                       STREAM RECONNAISSANCE                               │
│                     Revised for Bridge Inspection                         │
│                        Based on Thorne (1998)                             │
│                                                                           │
│                     SECTION 1—SITE DESCRIPTION                            │
│                                                                           │
│  ROAD NAME/NUMBER                            DATE                         │
│                                                                           │
│  BRIDGE NUMBER                                                            │
│                                                                           │
│  STREAM NAME                                                              │
│                                                                           │
│  GPS COORDINATES                                                          │
│                                                                           │
│                                                                           │
│              SECTION 2—REGION AND VALLEY DESCRIPTION                      │
└─────────────────────────────────────────────────────────────────────────┘
```

STREAM RECONNAISSANCE
Revised for Bridge Inspection
Based on Thorne (1998)

SECTION 1—SITE DESCRIPTION

ROAD NAME/NUMBER   DATE

BRIDGE NUMBER

STREAM NAME

GPS COORDINATES

SECTION 2—REGION AND VALLEY DESCRIPTION

PART 1: WATERSHED

Land Use	Vegetation
☐ Natural	☐ None
☐ Agricultural	☐ Grass
☐ Urban	☐ Pasture
☐ Suburban	☐ Crops
☐ Rural	☐ Shrubs
☐ Industrial	☐ Deciduous Forest/trees
☐ Cattle grazing	☐ Coniferous Forest/trees

PART 2: RIVER VALLEY CONDITION

Valley Side Failures	Failure Locations
☐ None	☐ None
☐ Occasional	☐ Away from river
☐ Frequent	☐ Along river

PART 3: FLOOD PLAIN

Flood Plain Width	Land Use	Vegetation	Riparian Buffer Strip
☐ None	☐ Natural	☐ None	☐ None
☐ < 1 river width	☐ Agricultural	☐ Grass	☐ < 1 river width
☐ 1–5 river widths	☐ Urban	☐ Pasture	☐ 1–5 river widths
☐ 5–10 river widths	☐ Suburban	☐ Orchards	☐ > 5 river widths
☐ > 10 river widths	☐ Rural	☐ Crops	
	☐ Industrial	☐ Shrubs	
	☐ Mining	☐ Deciduous Forest/trees	
	☐ Cattle grazing	☐ Coniferous Forest/trees	

Figure 18. Simplified and revised reconnaissance sheets based on Thorne, sheet 1.[2]

PART 4: VERTICAL CONFINEMENT

Terraces	Levees	Levee Location
☐ None	☐ None	☐ Along channel bank
☐ Left bank	☐ Natural	☐ Setback < 1 river width
☐ Right bank	☐ Constructed	☐ Setback > 1 river width

PART 5: LATERAL RELATION OF CHANNEL TO VALLEY

Planform	Meander Characteristics
☐ Straight	☐ Mild bends
☐ Meandering	☐ Moderate bends
☐ Braided	☐ Tight bends
☐ Anastomosed	
☐ Engineered	

SECTION 3—CHANNEL DESCRIPTION

PART 6: CHANNEL DESCRIPTION (select all that apply)

Bed Controls	Control Types	Width Controls	Control Types	Other
☐ None	☐ None	☐ None	☐ None	☐ Debris
☐ Occasional	☐ Bedrock	☐ Occasional	☐ Bedrock	☐ Mining
☐ Frequent	☐ Boulders	☐ Frequent	☐ Boulders	☐ Reservoir
☐ Confined	☐ Gravel armor	☐ Confined	☐ Gravel armor	☐ Knickpoint
	☐ Bridge protection		☐ Bridge protection	
	☐ Grade control		☐ Bridge abutments	
	☐ Debris		☐ Bank stabilization	
	☐ Dams (beaver, engineered)		☐ Debris	

Flow Habit   **Channel Width = _____**

☐ Perennial

☐ Flashy perennial   **M-B Classification**   **Corps Classification (Other)**

☐ Intermittent   ☐ Cascade or step-pool   ☐ Modified (engineered)

☐ Ephemeral   ☐ Plane, pool-riffle, dune-ripple   ☐ Regulated

☐ Braided   ☐ Arroyo

PART 7: BED SEDIMENT DESCRIPTION (select all that apply)

Bed Material	Bar Types	Bar Material	Bar Vegetation	Bar Width
☐ Clay	☐ None	☐ Silt	☐ None	☐ None
☐ Silt	☐ Alternate bars	☐ Sand	☐ Grasses	☐ Narrow
☐ Sand	☐ Point bars	☐ Gravel	☐ Reeds/shrubs	☐ Moderate
☐ Gravel	☐ Midchannel bars	☐ Cobbles	☐ Trees	☐ Wide
☐ Cobbles	☐ Diagonal bars			
☐ Boulders	☐ Irregular/combination			
☐ Bedrock	☐ Braided	**Percent Sand in Bed = _____ %**		

Figure 19. Simplified and revised reconnaissance sheets based on Thorne, sheet 2.[2]

SECTION 4—BANK SURVEY (select all that apply)		
Bank Characteristic	**Left Bank**	**Right Bank**
Bank material	☐ Clay ☐ Silt ☐ Sand ☐ Gravel ☐ Cobbles ☐ Boulders ☐ Bedrock	☐ Clay ☐ Silt ☐ Sand ☐ Gravel ☐ Cobbles ☐ Boulders ☐ Bedrock
Layer material	☐ No layers ☐ Cohesive ☐ Sand ☐ Gravel ☐ Cobbles ☐ Boulders	☐ No layers ☐ Cohesive ☐ Sand ☐ Gravel ☐ Cobbles ☐ Boulders
Bank height		
Bank slope	☐ Steep ☐ Moderate ☐ Shallow	☐ Steep ☐ Moderate ☐ Shallow
Bank vegetation	☐ None ☐ Grasses/annuals ☐ Reeds/shrubs ☐ Trees Falling trees? ☐ Yes ☐ No Tree density: ☐ Sparse ☐ Dense Tree health: ☐ Good ☐ Poor Tree ages: ☐ Young ☐ Mature ☐ Old Tree diversity? ☐ Yes ☐ No	☐ None ☐ Grasses/annuals ☐ Reeds/shrubs ☐ Trees Falling trees? ☐ Yes ☐ No Tree density: ☐ Sparse ☐ Dense Tree health: ☐ Good ☐ Poor Tree ages: ☐ Young ☐ Mature ☐ Old Tree diversity? ☐ Yes ☐ No
Bank erosion and failure location	Location of erosion: 　☐ Outside meander bend 　☐ Inside meander bend 　☐ Opposite bar or obstruction 　☐ General Type of erosion: 　☐ Fluvial 　☐ Geotechnical	Location of erosion: 　☐ Outside meander bend 　☐ Inside meander bend 　☐ Opposite bar or obstruction 　☐ General Type of erosion: 　☐ Fluvial 　☐ Geotechnical

Figure 20. Simplified and revised reconnaissance sheets based on Thorne, sheet 3.[2]

6. EXAMPLES

This chapter provides two examples for using the stream stability assessment method based on photos shown in the appendix A. The first example is Jayne Avenue over Jacalitos Creek near Coalinga, CA. Using http://terraserver-usa.com and the coordinates in table 5, a larger view of the bridge-stream intersection can be seen. The stream is seen to be mobile laterally, as evidenced by the large scars, deposits, and remnant channels. The photos of Jacalitos Creek in appendix A show views of the channel upstream and downstream of the bridge. Using the modified Thorne reconnaissance sheets given in figures 18–20, data collection begins by recording the map location, the GPS location, and the date. Next, characteristics of the watershed and flood plain are recorded. From the aerial photo on http://terraserver-usa.com, it is clear that the watershed use is primarily agricultural, while at least a portion of the flood plain is natural to allow for lateral movement of the stream. A visit to the site showed that cattle grazing is also a large part of the land use. The channel is braided to meandering with a riparian buffer of trees and shrubs upstream, but minimal buffer downstream. The channel was classified according to the Montgomery-Buffington and USACE methods. This yields a dune-ripple bed that is meandering to braided. The width and depth measurements are taken upstream and downstream of the bridge at approximate bankfull elevation. (Note that these measurements are not needed for the assessment method, but provide a record from which to compare the channel over many years.) The number of measurements needed depends on the variability of the channel dimensions. For this channel, three measurements upstream and downstream of the bridge were adequate to describe an average width-to-depth ratio. The measurements should be taken out of the influence of the bridge so that contraction scour does not influence the recorded dimensions. The general rule of thumb for reach length is 20 channel widths. In this case, where the channel is fairly uniform, 10 widths upstream and 10 downstream are adequate. The remaining data are recorded for the channel sediment, obstructions, and bank characteristics. The primary purpose of recording these data on the modified Thorne sheets is to help familiarize the user with the channel, focus the user's attention on various aspects of the channel, and provide a record of conditions that can be compared in subsequent years. While still onsite and after the reconnaissance sheets are filled out, the stability assessment sheet in table 8 should be used to determine the ratings for each of the 13 indicators. For Jacalitos Creek, the ratings and the total sum are given in table 9. Since this channel is in the dune-ripple category, table 10 is used to determine the overall rating.

The second example is S.R. 445 over Roaring Run in the Valley and Ridge Province of central Pennsylvania. Photos for this channel are provided in appendix A. Using http://terraserver-usa.com and the coordinates in table 5 to obtain aerial photos provides a larger view of the channel, the bridge-channel intersection, and watershed characteristics. Land use is primarily natural forest. The channel classifies as a step-pool channel according to the Montgomery-Buffington method and a mountain torrent, according to the USACE method. The modified Thorne sheets are completed to familiarize and focus the user on the channel, flood plain, and watershed. In this channel, the dimensions and other characteristics are relatively constant, so 10 channel widths upstream and downstream were adequate for estimating average dimensions and other observations. Cross-sectional width and depth were measured at the approximate bankfull elevation. Since the channel is very uniform both upstream and downstream, only one to two cross sections needed to be measured. The detail of the cross-sectional measurement depends on the user and the need to have detailed cross sections to compare in the future. For the purposes of the assessment method developed here, no cross-sectional data are used; however, as stated previously, it may be desirable to measure several cross sections in greater detail for future use. For the Thorne reconnaissance method, only an average width and depth are recorded. Although it is unnecessary for an initial or near-bridge channel stability assessment, it should be noted that walking much longer lengths of the channel can reveal disturbances that could eventually affect the bridge-channel intersection of interest. After the modified Thorne sheets are completed and the user is familiar with the stream, the stability assessment ratings can be determined. In this example, all indicators except for #13 show that the channel is very stable. The alignment of the channel and the bridge, however, cause #13 to be much higher than the others. The tight bend on which the bridge sits could clearly migrate in the future and cause problems at the bridge. Otherwise, the channel stability is rated as excellent, according to table 11.

If a stability rating is determined to be fair or poor, it might be desirable to return to the site for more detailed channel measurements by survey. Lateral movement and bed degradation can be measured over time by using such detailed, repeated measurements. However, to conduct rapid, preliminary assessments, as provided in this report, such detailed measurements are not necessary. If a channel is deemed to be stable (good to excellent), additional detailed measurements likely will not be needed. Thus, this method can be used as a decisionmaking tool regarding the need for more detailed and costly assessments.

7. CONCLUSIONS

The stream stability assessment method developed here provides a self-contained, preliminary assessment of channel stability conditions as they affect bridge foundations. This method provides a quick assessment of conditions for the purpose of judging whether a more extensive geomorphic study or complete hydraulics (HEC-20, Level II) and sediment transport analyses (HEC-20, Level III) are needed to assess further the potential for adverse conditions developing at the bridge. As such, the method assists in decisionmaking with respect to bridge design, repair, rehabilitation, or replacement.

Although the stability assessment method was developed for use in any physiographic region in the United States, it was not possible to make observations within every region. In addition, it was not possible to sample every stream type and stream order within each region. Therefore, the method should be used cautiously and with questioning so that the results represent the stability at the bridge-stream intersection.

The results of this project provide a simplified methodology for assessing channel stability at bridge-stream intersections. It is intended to be used by qualified bridge inspectors and hydraulic engineers. This method developed is self-contained and does not require any other data or formal method of data collection other than the descriptors given in table 8. However, field forms were developed based on Thorne to help observers focus attention on specific aspects of a stream, make consistent observations, and record the observations systematically.[2] A photo album of the bridges in this study is provided in appendix A.

APPENDIX A

This section contains a photo album for bridge-stream intersections according to physiographic regions.

Figure 21. Dry Creek, Pacific Coastal—upstream from bridge.

Figure 22. Dry Creek, Pacific Coastal—downstream from bridge.

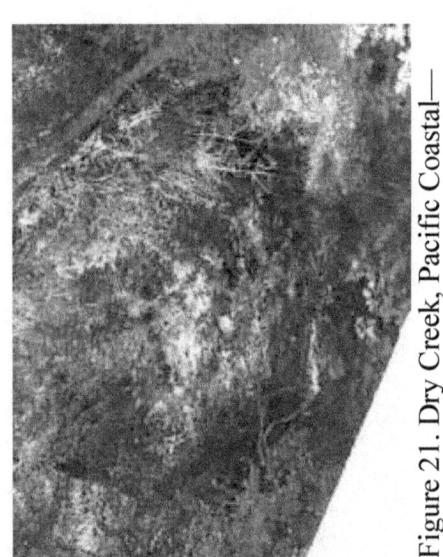

Figure 23. Dry Creek, Pacific Coastal—upstream under bridge, photo 1.

Figure 24. Dry Creek, Pacific Coastal—upstream under bridge, photo 2.

Figure 25. Dutch Bill Creek, Pacific Coastal—upstream from under bridge.

Figure 26. Dutch Bill Creek, Pacific Coastal—downstream at bridge.

Figure 28. Dutch Bill Creek, Pacific Coastal—upstream through bridge.

Figure 27. Dutch Bill Creek, Pacific Coastal—downstream from under bridge.

87

Figure 29. Buena Vista Creek, Pacific Coastal—upstream from bridge.

Figure 30. Buena Vista Creek, Pacific Coastal—downstream from bridge.

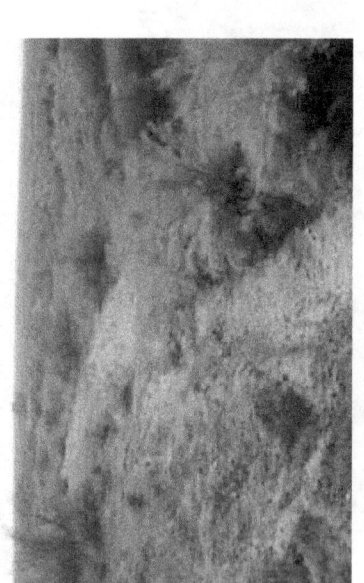

Figure 31. Buena Vista Creek, Pacific Coastal—downstream under bridge.

Figure 32. Buena Vista Creek, Pacific Coastal—upstream from under bridge.

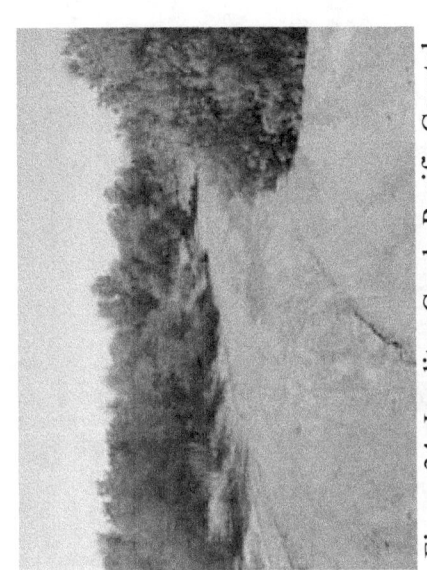

Figure 34. Jacalitos Creek, Pacific Coastal—upstream from bridge.

Figure 36. Jacalitos Creek, Pacific Coastal—downstream from under bridge.

Figure 33. Jacalitos Creek, Pacific Coastal—downstream from bridge.

Figure 35. Jacalitos Creek, Pacific Coastal—looking downstream at bridge.

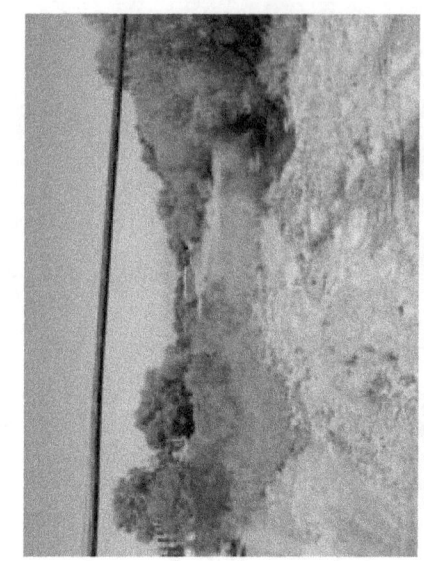

Figure 38. Murietta Creek, Pacific Coastal—upstream from bridge.

Figure 40. Murietta Creek, Pacific Coastal—looking upstream at bridge.

Figure 37. Murietta Creek, Pacific Coastal—downstream from bridge.

Figure 39. Murietta Creek, Pacific Coastal—upstream toward bridge.

90

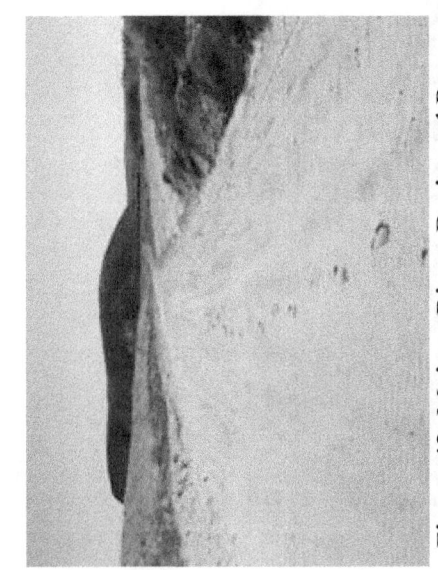

Figure 41. Mojave River, Basin and Range—
upstream from bridge, photo 1.

Figure 42. Mojave River, Basin and Range—
upstream from bridge, photo 2.

Figure 43. Mojave River, Basin and Range—
looking downstream at bridge.

Figure 44. Mojave River, Basin and Range—
downstream from bridge.

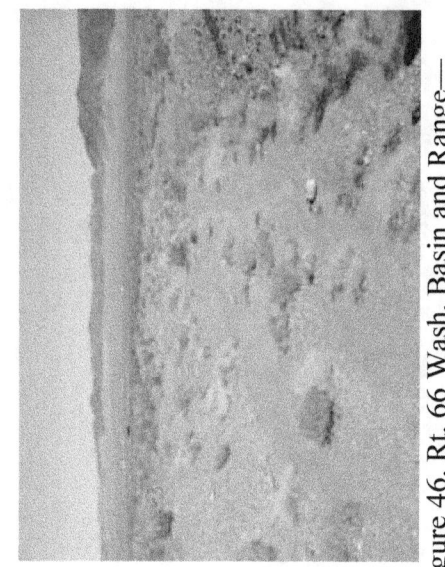

Figure 46. Rt. 66 Wash, Basin and Range—downstream from bridge.

Figure 48. Rt. 66 Wash, Basin and Range—looking upstream at bridge.

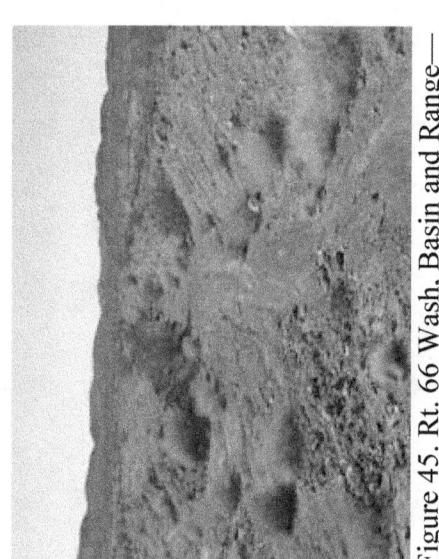

Figure 45. Rt. 66 Wash, Basin and Range—upstream from bridge.

Figure 47. Rt. 66 Wash, Basin and Range—looking downstream at bridge.

Figure 50. Sacramento Wash, Basin and Range—
downstream from bridge.

Figure 52. Sacramento Wash, Basin and Range—
looking downstream at bridge.

Figure 49. Sacramento Wash, Basin and Range—
upstream under bridge.

Figure 51. Sacramento Wash, Basin and Range—
downstream under bridge.

Figure 54. Rio Puerco, Trans Pecos—upstream from bridge.

Figure 56. Rio Puerco, Trans Pecos—looking downstream at bridge.

Figure 53. Rio Puerco, Trans Pecos—downstream from bridge.

Figure 55. Rio Puerco, Trans Pecos—looking upstream at bridge.

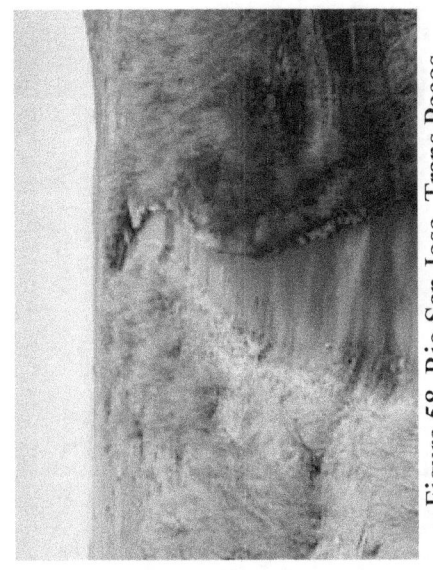

Figure 58. Rio San Jose, Trans Pecos—downstream from bridge.

Figure 60. Rio San Jose, Trans Pecos—looking downstream at bridge.

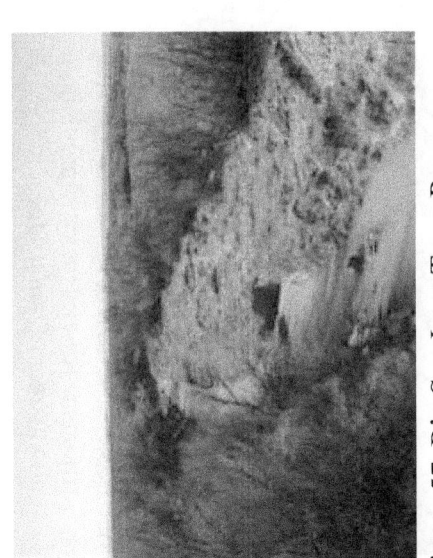

Figure 57. Rio San Jose, Trans Pecos—upstream from bridge.

Figure 59. Rio San Jose, Trans Pecos—looking upstream at bridge.

Figure 62. Arkansas River, Rocky Mountains—upstream from bridge.

Figure 64. Arkansas River, Rocky Mountains—bridge #2 downstream from other bridge.

Figure 61. Arkansas River, Rocky Mountains—looking downstream at bridge.

Figure 63. Arkansas River, Rocky Mountains—downstream from bridge.

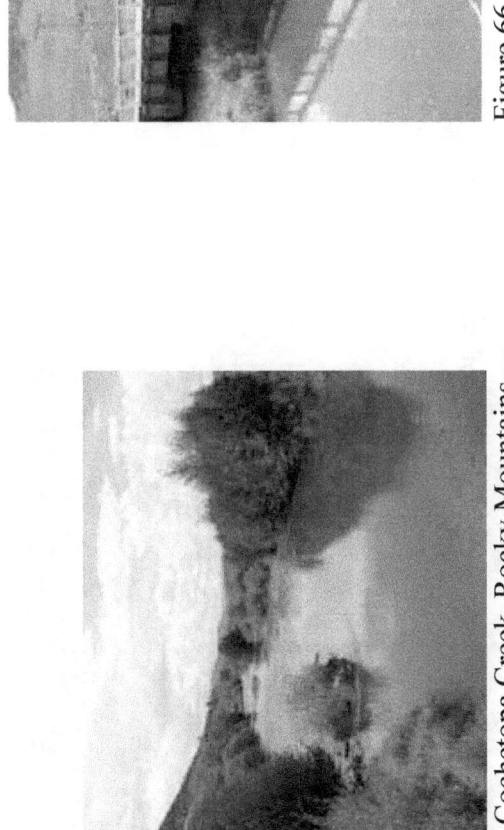

Figure 66. Cochetopa Creek, Rocky Mountains—looking downstream at bridge.

Figure 65. Cochetopa Creek, Rocky Mountains—downstream from bridge.

Figure 67. Cochetopa Creek, Rocky Mountains—upstream from bridge.

97

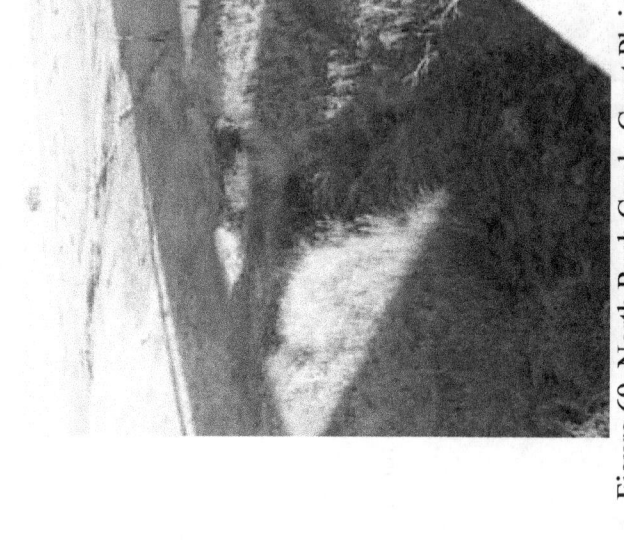

Figure 69. North Rush Creek, Great Plains—upstream from bridge.

Figure 68. North Rush Creek, Great Plains—upstream of bridge.

Figure 70. North Rush Creek, Great Plains—downstream from bridge.

Figure 72. Saline River, Great Plains—downstream from bridge.

Figure 73. Saline River, Great Plains—looking downstream at bridge.

Figure 71. Saline River, Great Plains—upstream under bridge.

99

Figure 74. South Fork Solomon River, Great Plains—looking downstream at bridge, photo 1.

Figure 75. South Fork Solomon River, Great Plains—looking downstream at bridge, photo 2.

Figure 76. South Fork Solomon River, Great Plains—left bank.

Figure 77. South Fork Solomon River, Great Plains—downstream.

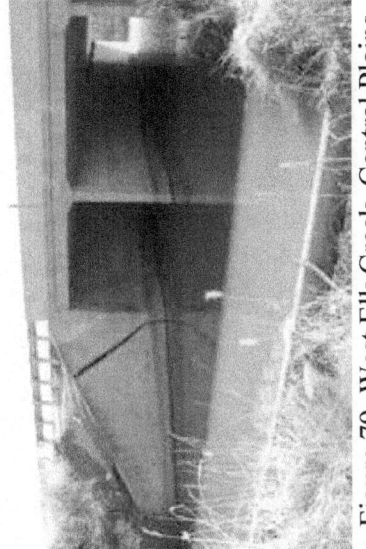

Figure 79. West Elk Creek, Central Plains— looking downstream at bridge, photo 2.

Figure 81. West Elk Creek, Central Plains— downstream from bridge.

Figure 78. West Elk Creek, Central Plains— looking downstream at bridge, photo 1.

Figure 80. West Elk Creek, Central Plains— upstream from bridge.

Figure 83. Beaver Creek, Central Plains—downstream from bridge.

Figure 85. Beaver Creek, Central Plains—facing downstream under bridge.

Figure 82. Beaver Creek, Central Plains—upstream from bridge.

Figure 84. Beaver Creek, Central Plains—facing upstream under bridge.

Figure 87. Brush Creek, Central Plains—downstream from bridge.

Figure 88. Brush Creek, Central Plains—downstream under bridge.

Figure 86. Brush Creek, Central Plains—upstream from bridge.

103

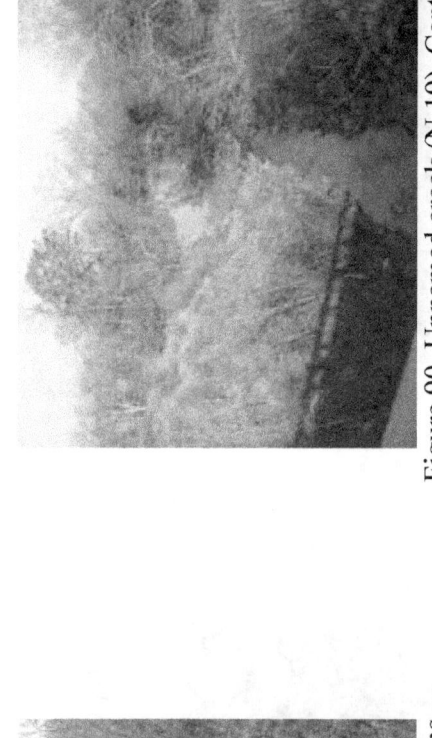

Figure 90. Unnamed creek (N 19), Central Plains—upstream from bridge.

Figure 92. Unnamed creek (N 19), Central Plains—downstream under bridge.

Figure 89. Unnamed creek (N 19), Central Plains—downstream from bridge.

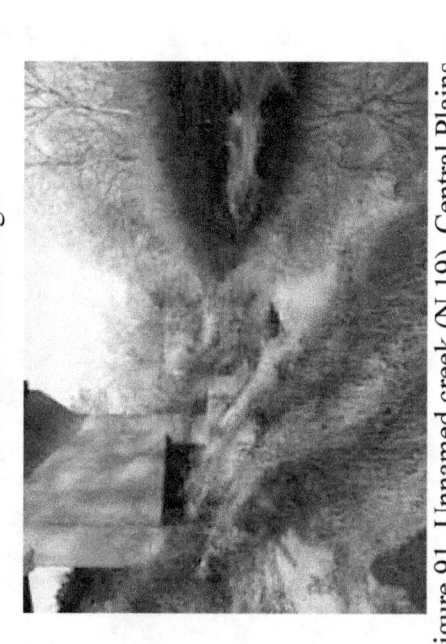

Figure 91. Unnamed creek (N 19), Central Plains—upstream under bridge.

104

Figure 94. East Fork, Interior Low Plateau—looking downstream at second bridge.

Figure 96. East Fork, Interior Low Plateau—looking downstream at bridge.

Figure 93. East Fork, Interior Low Plateau—upstream from bridge.

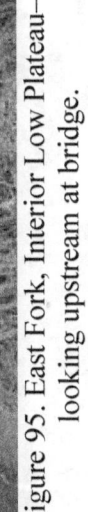

Figure 95. East Fork, Interior Low Plateau—looking upstream at bridge.

Figure 98. Honey Run, Interior Low Plateau—
downstream from bridge.

Figure 100. Honey Run, Interior Low Plateau—
looking downstream at bridge.

Figure 97. Honey Run, Interior Low Plateau—
upstream from bridge.

Figure 99. Honey Run, Interior Low Plateau—
looking upstream at bridge.

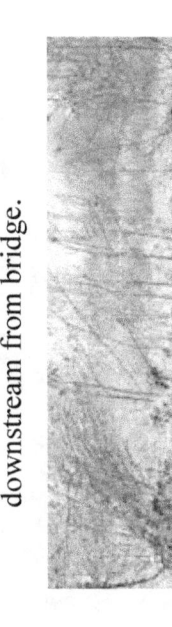

Figure 102. Unnamed creek (N 28), Interior Low Plateau—downstream from bridge.

Figure 104. Unnamed creek (N 28), Interior Low Plateau—rocky bank material.

Figure 101. Unnamed creek (N 28), Interior Low Plateau—upstream from bridge.

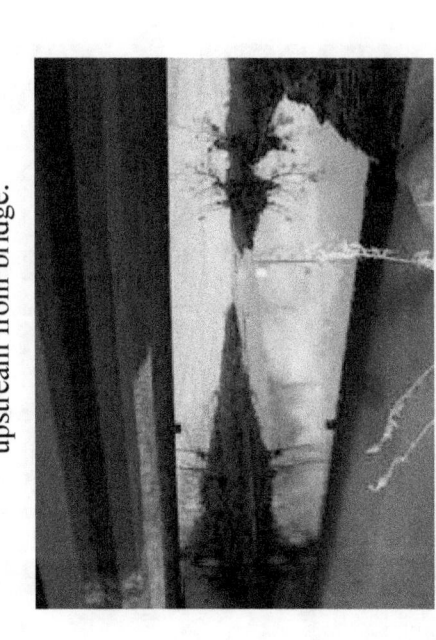

Figure 103. Unnamed creek (N 28), Interior Low Plateau—downstream under bridge.

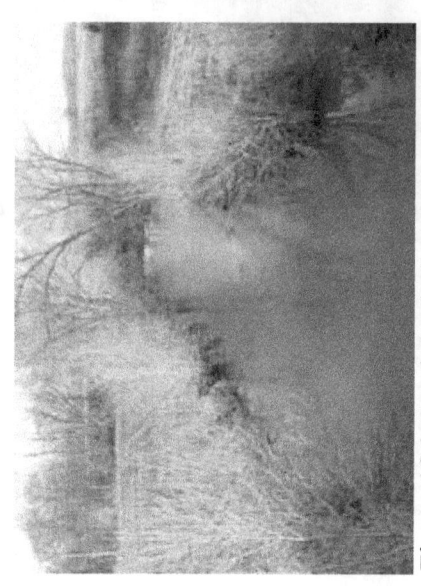

Figure 106. South Fork, Interior Low Plateau—upstream from bridge.

Figure 108. South Fork, Interior Low Plateau—looking downstream at bridge.

Figure 105. South Fork, Interior Low Plateau—downstream from bridge.

Figure 107. South Fork, Interior Low Plateau—looking upstream at bridge.

108

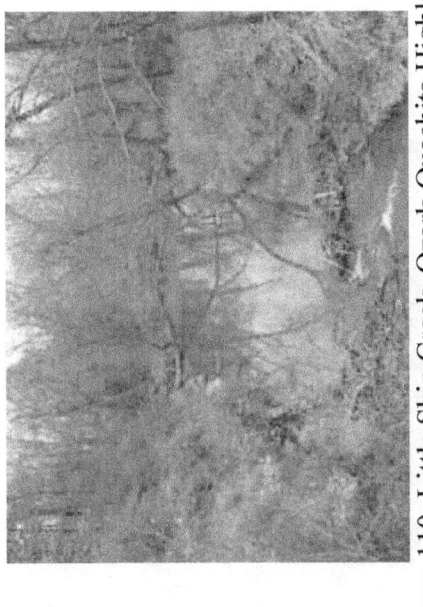

Figure 110. Little Skin Creek, Ozark-Ouachita Highlands—upstream from bridge.

Figure 112. Little Skin Creek, Ozark-Ouachita Highlands—looking downstream at bridge (right).

Figure 109. Little Skin Creek, Ozark-Ouachita Highlands—downstream from bridge.

Figure 111. Little Skin Creek, Ozark-Ouachita Highlands—looking downstream at bridge (left).

Figure 113. Unnamed creek (N 21), Ozark-Ouachita Highlands—downstream from bridge.

Figure 114. Unnamed creek (N 21), Ozark-Ouachita Highlands—upstream from bridge.

Figure 115. Unnamed creek (N 21), Ozark-Ouachita Highlands—looking downstream at bridge.

Figure 116. Unnamed creek (N 21), Ozark-Ouachita Highlands—looking upstream at bridge.

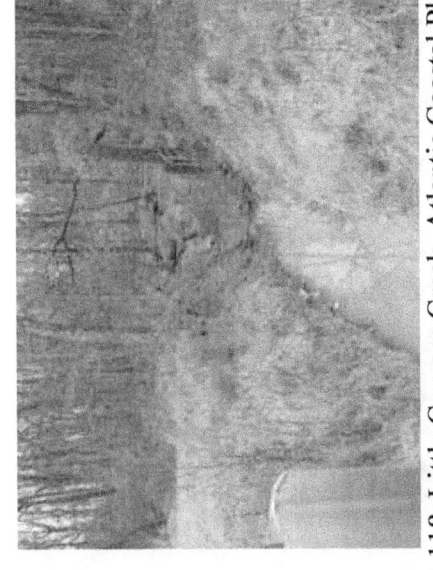

Figure 117. Little Cypress Creek, Atlantic Coastal Plain—
downstream from bridge.

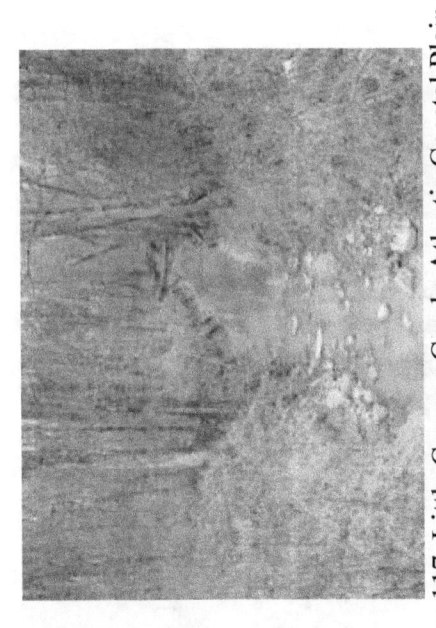

Figure 118. Little Cypress Creek, Atlantic Coastal Plain—
upstream from bridge.

Figure 119. Little Cypress Creek, Atlantic Coastal Plain—
looking downstream at bridge.

Figure 120. Little Cypress Creek, Atlantic Coastal Plain—
looking upstream at bridge.

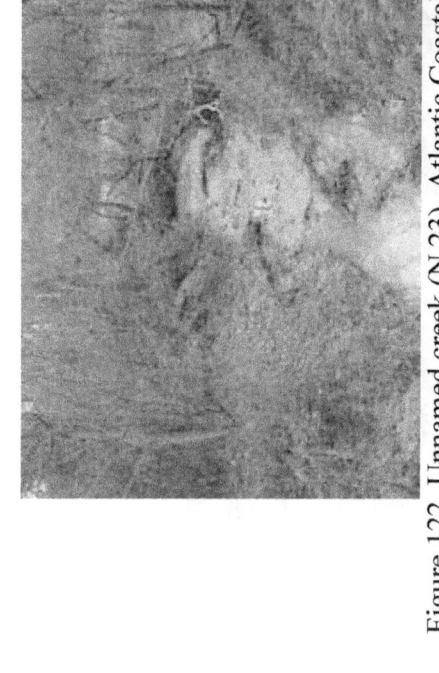

Figure 122. Unnamed creek (N 23), Atlantic Coastal Plain—upstream from bridge.

Figure 124. Unnamed creek (N 23), Atlantic Coastal Plain—looking upstream at bridge.

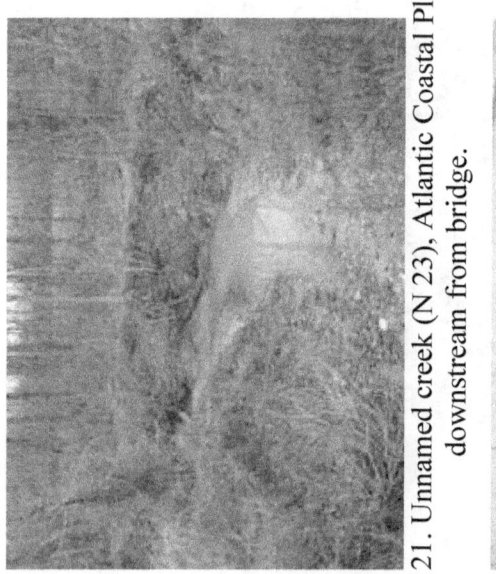

Figure 121. Unnamed creek (N 23), Atlantic Coastal Plain—downstream from bridge.

Figure 123. Unnamed creek (N 23), Atlantic Coastal Plain—looking downstream at bridge.

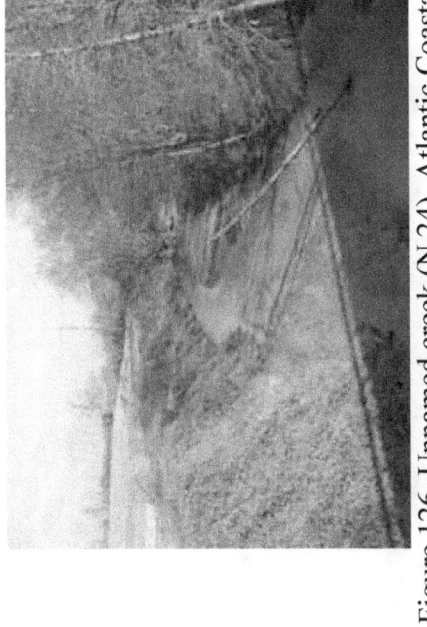

Figure 126. Unnamed creek (N 24), Atlantic Coastal Plain—downstream from bridge.

Figure 128. Unnamed creek (N 24), Atlantic Coastal Plain—looking downstream at bridge.

Figure 125. Unnamed creek (N 24), Atlantic Coastal Plain—upstream from bridge.

Figure 127. Unnamed creek (N 24), Atlantic Coastal Plain—looking upstream at bridge.

113

Figure 130. Peace River, Atlantic Coastal Plain—looking downstream at bridge.

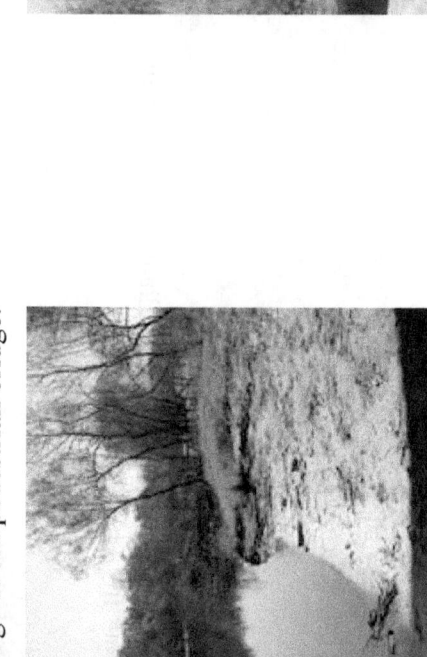

Figure 132. Peace River, Atlantic Coastal Plain—upstream from old pedestrian bridge.

Figure 129. Peace River, Atlantic Coastal Plain—upstream from bridge at old pedestrian bridge.

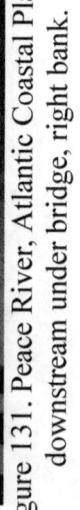

Figure 131. Peace River, Atlantic Coastal Plain—downstream under bridge, right bank.

Figure 133. Alligator Creek, Atlantic Coastal Plain—downstream from bridge.

Figure 135. Alligator Creek, Atlantic Coastal Plain—looking upstream at bridge.

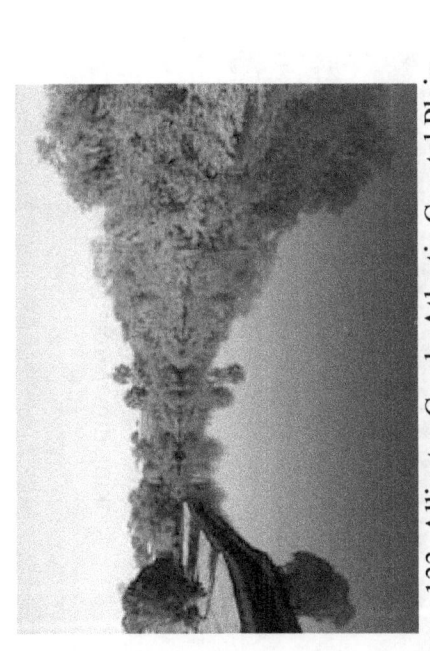

Figure 134. Alligator Creek, Atlantic Coastal Plain—looking downstream at bridge.

Figure 136. Alligator Creek, Atlantic Coastal Plain—upstream from bridge.

Figure 138. Stocketts Run, Atlantic Coastal Plain—downstream from bridge.

Figure 140. Stocketts Run, Atlantic Coastal Plain—looking upstream at bridge.

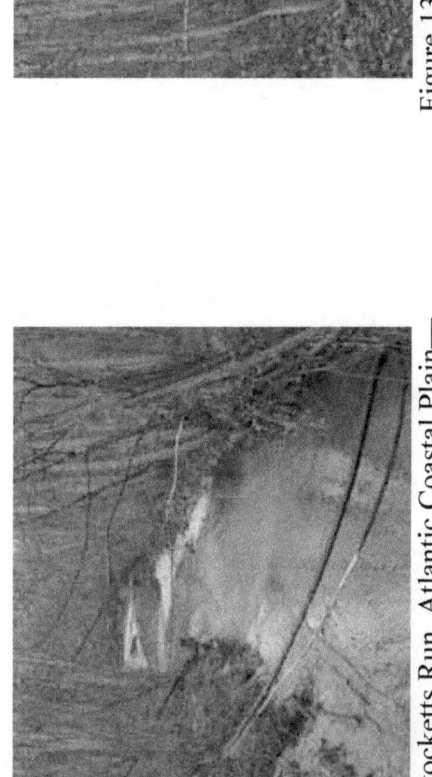

Figure 137. Stocketts Run, Atlantic Coastal Plain—upstream from bridge.

Figure 139. Stocketts Run, Atlantic Coastal Plain—looking downstream at bridge.

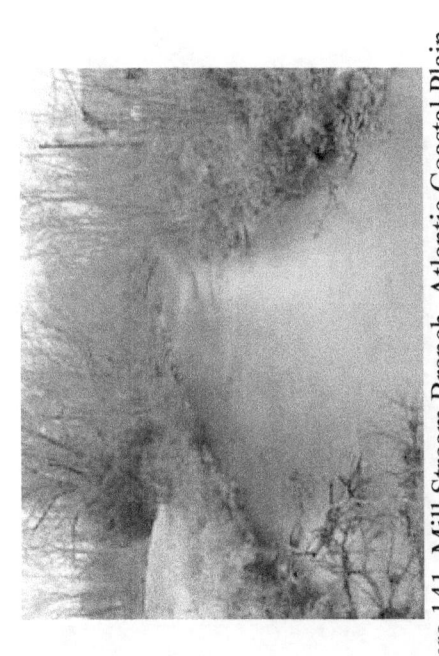

Figure 141. Mill Stream Branch, Atlantic Coastal Plain—upstream from bridge.

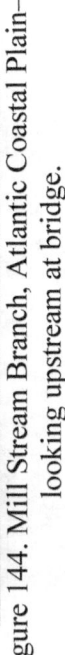

Figure 142. Mill Stream Branch, Atlantic Coastal Plain—downstream from bridge.

Figure 143. Mill Stream Branch, Atlantic Coastal Plain—looking downstream at bridge.

Figure 144. Mill Stream Branch, Atlantic Coastal Plain—looking upstream at bridge.

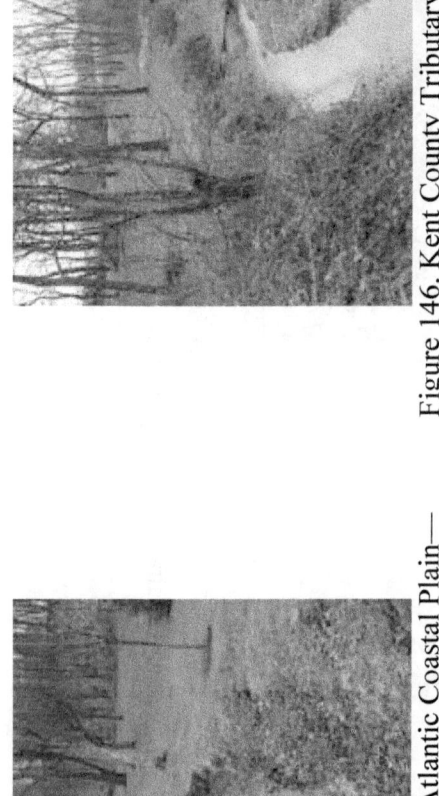

Figure 146. Kent County Tributary, Atlantic Coastal Plain— upstream from bridge.

Figure 148. Kent County Tributary, Atlantic Coastal Plain— looking upstream at bridge.

Figure 145. Kent County Tributary, Atlantic Coastal Plain— downstream from bridge.

Figure 147. Kent County Tributary, Atlantic Coastal Plain— looking downstream at bridge.

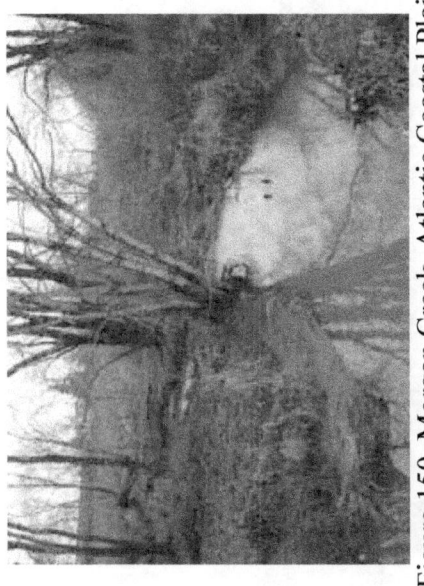

Figure 149. Morgan Creek, Atlantic Coastal Plain—upstream from bridge.

Figure 151. Morgan Creek, Atlantic Coastal Plain—looking upstream at bridge.

Figure 150. Morgan Creek, Atlantic Coastal Plain—downstream from bridge.

Figure 152. Morgan Creek, Atlantic Coastal Plain—looking downstream at bridge.

Figure 153. Hammond Branch, Atlantic Coastal Plain—upstream from bridge.

Figure 154. Hammond Branch, Atlantic Coastal Plain—downstream from bridge.

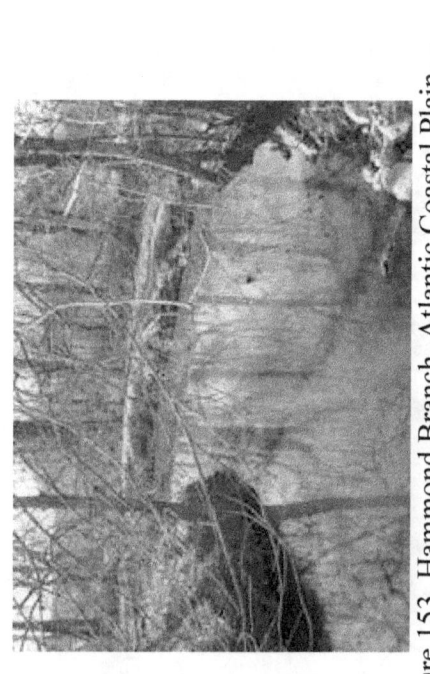

Figure 155. Hammond Branch, Atlantic Coastal Plain—looking downstream at bridge.

Figure 156. Hammond Branch, Atlantic Coastal Plain—looking upstream at bridge.

Figure 158. Pootatuck River, New England—upstream from bridge.

Figure 160. Pootatuck River, New England—looking upstream at bridge.

Figure 157. Pootatuck River, New England—downstream from bridge.

Figure 159. Pootatuck River, New England—looking downstream at bridge.

Figure 162. Mill River, New England—downstream from bridge.

Figure 164. Mill River, New England—looking downstream at bridge.

Figure 161. Mill River, New England—upstream from bridge.

Figure 163. Mill River, New England—looking upstream at bridge.

Figure 165. Aspetuck River, New England—upstream from bridge.

Figure 166. Aspetuck River, New England—downstream from bridge.

Figure 167. Aspetuck River, New England—looking upstream at bridge.

Figure 168. Aspetuck River, New England—looking downstream at bridge.

Figure 169. West Branch Saugatuck River, New England—downstream from bridge.

Figure 170. West Branch Saugatuck River, New England—upstream from bridge.

Figure 171. West Branch Saugatuck River, New England—looking downstream at bridge (bridge in foreground is the pedestrian bridge).

Figure 172. West Branch Saugatuck River, New England—looking upstream at bridge.

Figure 174. Mianus River, New England—downstream from bridge. Note weir.

Figure 176. Mianus River, New England—looking upstream at bridge.

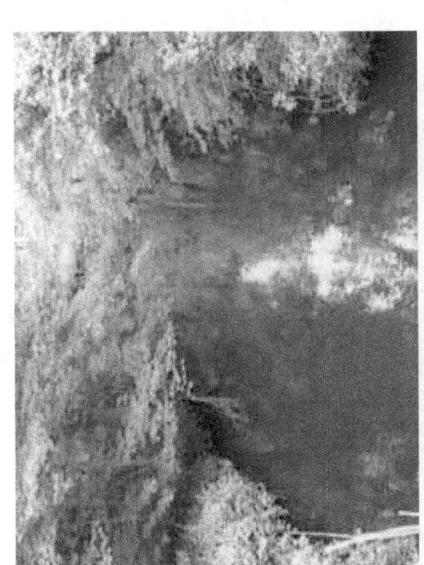

Figure 173. Mianus River, New England—upstream from bridge.

Figure 175. Mianus River, New England—looking downstream at bridge.

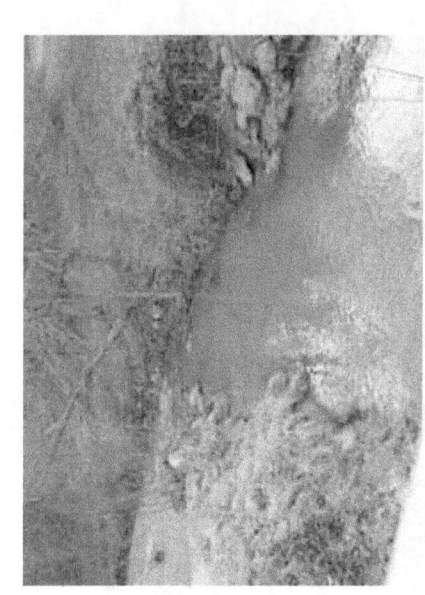

Figure 178. McKnown Creek, Appalachian Plateau—downstream from bridge.

Figure 180. McKnown Creek, Appalachian Plateau—looking upstream at bridge.

Figure 177. McKnown Creek, Appalachian Plateau—upstream from bridge.

Figure 179. McKnown Creek, Appalachian Plateau—looking downstream at bridge.

Figure 181. Wolf Run, Appalachian Plateau—upstream from bridge.

Figure 182. Wolf Run, Appalachian Plateau—downstream from bridge.

Figure 183. Wolf Run, Appalachian Plateau—looking upstream at bridge.

Figure 184. Wolf Run, Appalachian Plateau—upstream face of bridge.

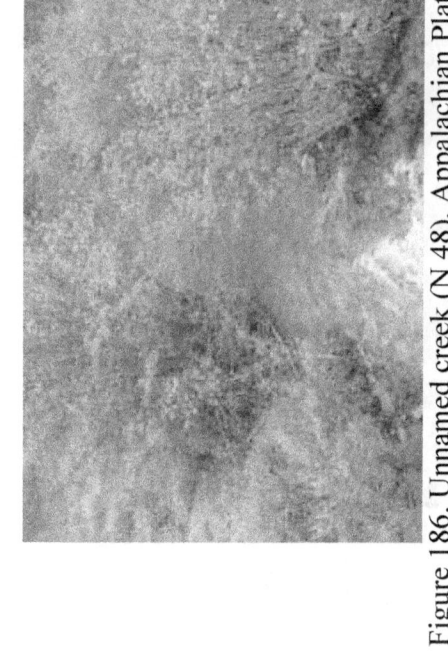

Figure 185. Unnamed creek (N 48), Appalachian Plateau—upstream from bridge.

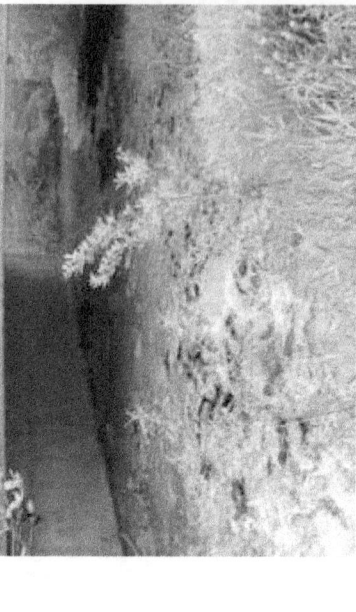

Figure 186. Unnamed creek (N 48), Appalachian Plateau—downstream from bridge.

Figure 187. Unnamed creek (N 48), Appalachian Plateau—looking downstream at bridge.

Figure 188. Unnamed creek (N 48), Appalachian Plateau—looking upstream through bridge.

Figure 190. Reids Run, Appalachian Plateau—downstream from bridge.

Figure 192. Reids Run, Appalachian Plateau—looking upstream at bridge.

Figure 189. Reids Run, Appalachian Plateau—upstream from bridge.

Figure 191. Reids Run, Appalachian Plateau—looking downstream at bridge.

129

Figure 194. Piney Creek, Appalachian Plateau—upstream from bridge.

Figure 196. Piney Creek, Appalachian Plateau—looking upstream at bridge.

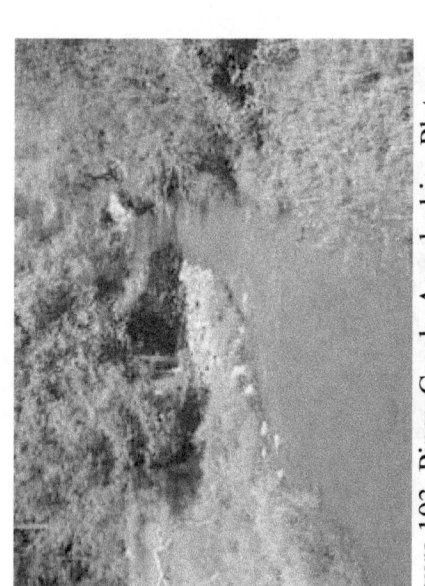

Figure 193. Piney Creek, Appalachian Plateau—downstream from bridge.

Figure 195. Piney Creek, Appalachian Plateau—looking downstream at bridge.

Figure 198. Sandy Creek, Appalachian Plateau—downstream from bridge.

Figure 200. Sandy Creek, Appalachian Plateau—looking upstream at bridge.

Figure 197. Sandy Creek, Appalachian Plateau—upstream from bridge.

Figure 199. Sandy Creek, Appalachian Plateau—looking downstream at bridge.

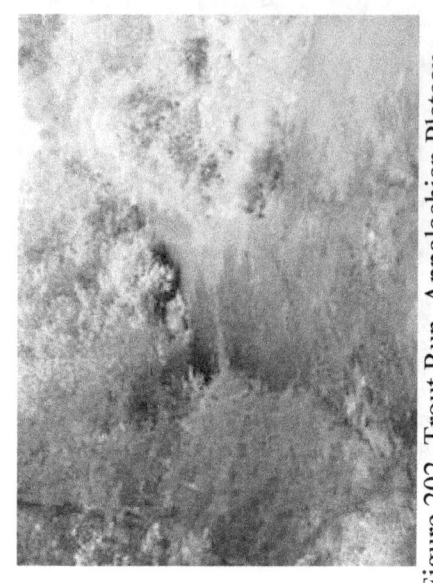

Figure 202. Trout Run, Appalachian Plateau—downstream from bridge.

Figure 204. Trout Run, Appalachian Plateau—upstream face of bridge.

Figure 201. Trout Run, Appalachian Plateau—upstream from bridge.

Figure 203. Trout Run, Appalachian Plateau—looking upstream at bridge.

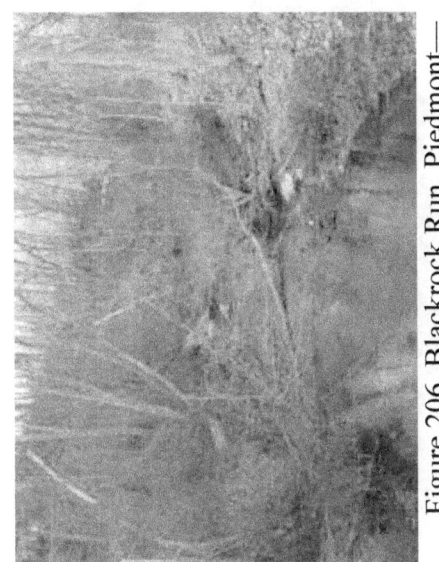

Figure 206. Blackrock Run, Piedmont—
downstream from bridge.

Figure 208. Blackrock Run, Piedmont—
looking upstream at bridge.

Figure 205. Blackrock Run, Piedmont—
upstream from bridge.

Figure 207. Blackrock Run, Piedmont—
looking downstream at bridge.

133

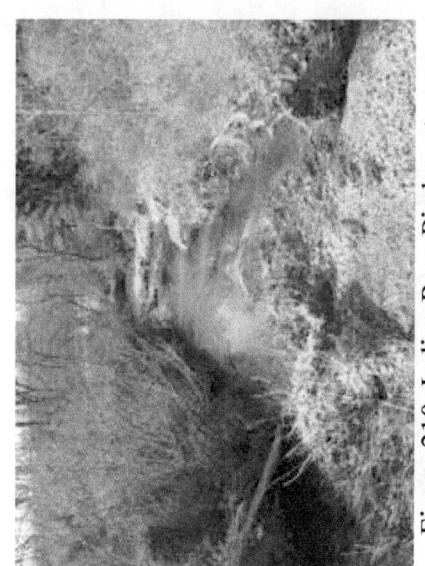

Figure 210. Indian Run, Piedmont— downstream from bridge.

Figure 212. Indian Run, Piedmont— looking upstream at bridge.

Figure 209. Indian Run, Piedmont— upstream from bridge.

Figure 211. Indian Run, Piedmont— looking downstream at bridge.

Figure 214. Middle Patuxent River, Piedmont—downstream from bridge.

Figure 216. Middle Patuxent River, Piedmont—looking upstream at bridge.

Figure 213. Middle Patuxent River, Piedmont—upstream from bridge.

Figure 215. Middle Patuxent River, Piedmont—looking downstream at bridge.

135

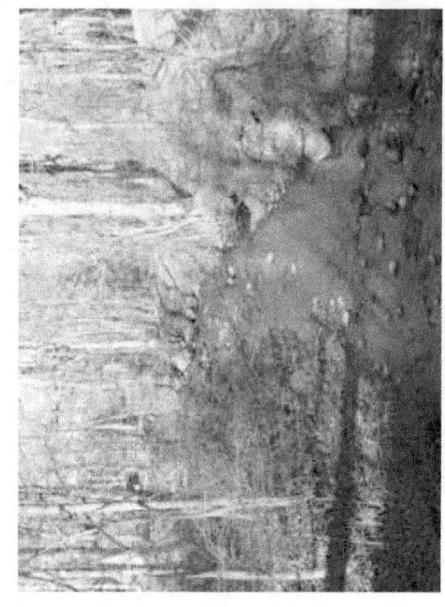

Figure 218. Atherton Tributary, Piedmont—downstream from bridge.

Figure 220. Atherton Tributary, Piedmont—looking downstream at bridge.

Figure 217. Atherton Tributary, Piedmont—upstream from bridge.

Figure 219. Atherton Tributary, Piedmont—looking upstream at bridge.

Figure 221. Little Elk Creek, Piedmont—upstream from bridge.

Figure 222. Little Elk Creek, Piedmont—downstream from bridge.

Figure 223. Little Elk Creek, Piedmont—looking upstream at bridge.

Figure 224. Little Elk Creek, Piedmont—looking downstream at bridge.

137

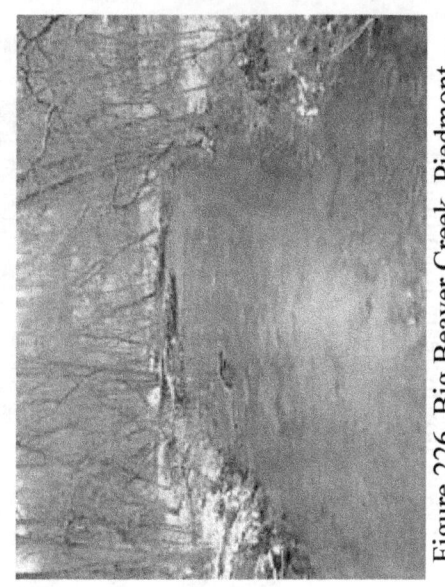

Figure 226. Big Beaver Creek, Piedmont—downstream from bridge.

Figure 228. Big Beaver Creek, Piedmont—looking upstream at bridge.

Figure 225. Big Beaver Creek, Piedmont—upstream from bridge.

Figure 227. Big Beaver Creek, Piedmont—looking downstream at bridge.

Figure 230. Buffalo Run, Valley and Ridge—downstream from bridge.

Figure 232. Buffalo Run, Valley and Ridge—looking upstream at bridge.

Figure 229. Buffalo Run, Valley and Ridge—upstream from bridge.

Figure 231. Buffalo Run, Valley and Ridge—looking downstream at bridge.

Figure 234. Roaring Run, Valley and Ridge—upstream from bridge.

Figure 236. Roaring Run, Valley and Ridge—looking upstream at bridge.

Figure 233. Roaring Run, Valley and Ridge—downstream from bridge.

Figure 235. Roaring Run, Valley and Ridge—looking downstream at bridge.

140

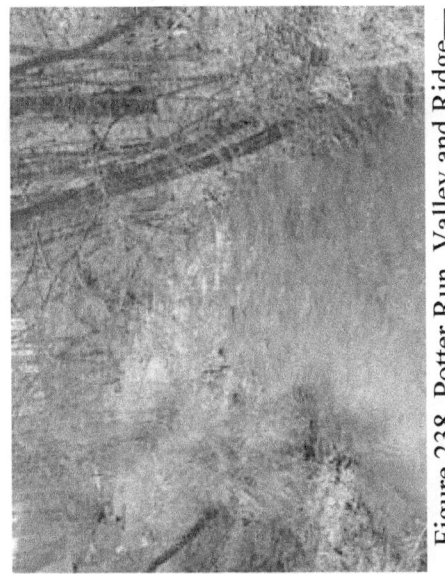

Figure 238. Potter Run, Valley and Ridge—upstream from bridge.

Figure 240. Potter Run, Valley and Ridge—looking upstream at bridge.

Figure 237. Potter Run, Valley and Ridge—downstream from bridge.

Figure 239. Potter Run, Valley and Ridge—looking downstream at bridge.

141

REFERENCES

1. Johnson, P.A., Gleason, G., and Hey, R.D. (1999). "Rapid assessment of channel stability in the vicinity of a road crossing." *Journal of Hydraulic Engineering*, American Society of Civil Engineers (ASCE), 125(6), 645–652.

2. Thorne, C.R. (1998). *Stream Reconnaissance Handbook: Geomorphological Investigation and Analysis of River Channels*. John Wiley & Sons Ltd., Chichester, England.

3. Lagasse, P.F., Schall, J.D., Johnson, F., Richardson, E.V., Richardson, J.R., and Chang, F. (2001). *Stream Stability at Highway Structures, Second Edition*. U.S. Department of Transportation, Report No. FHWA-IP-90-014, HEC-20-ED-2, FHWA, Washington, DC.

4. Richardson, E.V. and Davis, S.R. (2001). *Evaluating Scour at Bridges, Fourth Edition*. U.S. Department of Transportation, HEC-18, FHWA-IP-90-017. 132 pp.

5. Pfankuch, D.J. (1978). *Stream Reach Inventory and Channel Stability Evaluation*. Unpublished Report, U.S. Department of Agriculture, Forest Service, Northern Region, 25 pp.

6. Simon, A. and Downs, P.W. (1995). "An interdisciplinary approach to evaluation of potential instability in alluvial channels." *Geomorphology*, 12, 215–232.

7. Thorne, C.R., Allen, R.G., and Simon, A. (1996). "Geomorphological river channel reconnaissance for river analysis, engineering, and management." *Transactions of the Institute of British Geographer*, NS 21, 469–483.

8. Montgomery, D.R. and Buffington, J.M. (1993). *Channel Classification, Prediction of Channel Response, and Assessment of Channel Condition*. Washington State Department of Natural Resources Report TFW-SH10-93-002, Olympia, WA.

9. Montgomery, D.R. and Buffington, J.M. (1997). "Channel-Reach Morphology in Mountain Drainage Basins." *Geological Society of America Bulletin*, 109:596–611.

10. U.S. Army Corps of Engineers (1994). *Engineering and Design: Channel Stability Assessment for Flood Control Projects*. Engineer Manual No. 1110-2-1418, CECW-EH-D, Department of the Army, U.S. Army Corps of Engineers, Washington, DC.

11. Brookes, A. (1987). "River channel adjustments downstream from channelization works in England and Wales." *Earth Surface Processes and Landforms*, 12, 337–351.

12. Johnson, P.A. and Heil, T.M. (1996). "Uncertainty in estimating bankfull conditions." *Water Resources Bulletin*, 32(6), 1283–1292.

13. Chorley, R.J. and Kennedy, B.A. (1971). *Physical Geography: A Systems Approach*. Prentice-Hall, London, UK.

14. Richards, K. (1987). *River Channels: Environment and Process*. Institute of British Geographers, Special Publications Series, 17, Oxford, UK.

15. Knox, J.C. (1976). "Concept of the graded stream." *In Theories of Landform Development, Sixth Annual Geomorphology Symposia Series*. State University of New York, Binghamton, NY, 169–198.

16. Ritter, D.F. (1987). "Fluvial processes in the mountains and intermontane basins." In *Geomorphic Systems of North America* (W.L. Graf, Ed.) Geological Society of America, Boulder, CO, 220–228.

17. Lane, E.W. (1955). "The importance of fluvial geomorphology in hydraulic engineering." Proc. ASCE, 81, Paper 745, 1–17.

18. Hey, R.D. (1982). "Gravel-bed rivers: process and form." In Hey, R.D., Bathurst, J.C., and Thorne, C.R. (Eds.). *Gravel-Bed Rivers*. Wiley & Sons, Chichester, England, 5–13.

19. Richards, K.S. and Lane, S.N. (1997). "Prediction of morphological changes in unstable channels." In Thorne, C.R., Hey, R.D., and Newson, M.D. (Eds.). *Applied Fluvial Geomorphology for River Engineering and Management*. Wiley & Sons, New York, NY, Chapter 10.

20. Montgomery, D.R. and MacDonald, L.H. (2002). "Diagnostic approach to stream channel assessment and monitoring." *Journal of the American Water Resources Association*, 38(1), 1–16.

21. Ruhe, R.V. (1970). "Stream regimen and man's manipulation." In Coates, D.R., Ed., *Environmental Geomorphology—Annual Geomorphology Symposia, October 1970, Proceedings*, State University of New York, Binghamton, NY, 9–23.

22. Emerson, J.W. (1971). "Channelization, a case study." *Science*, 173, 325–326.

23. Wilson, D.V. (1979). "Changes in channel characteristics, 1938–1974, of the Homochitto River and tributaries, Mississippi." *U.S. Geological Survey Open File Report 79–554*, 18 pp.

24. Simon, A. (1989). "A model of channel response in disturbed alluvial channels." *Earth Surface Processes and Landforms*, 14, 11–26.

25. Simon, A. (1992). "Energy, time, and channel evolution in catastrophically disturbed fluvial systems." In J.D. Phillips and W.H. Renwick (Eds.) Geomorphic Systems. *Journal of Geomorphology*, 5, 345–372.

26. Heil, T. and Johnson, P.A. (1996). "Assessment and implications of local channel instability on the prediction of bridge scour." *Proceedings of the North American Water and Environment Congress 1996*, ASCE, Anaheim, CA.

27. Johnson, P.A. and Simon, A. (1997). "Reliability of bridge foundations in modified channels." *Journal of Hydraulic Engineering*, ASCE, 123(7), 648–651.

28. Johnson, P.A., Tereska, R.L., and Brown, E.R. (2002). "Using technical adaptive management to improve design guidelines for urban in-stream structures." *Journal of the American Water Resources Association*, 38(4), 1143–1152.

29. Mitchell, P. (1990). *The Environmental Condition of Victorian Streams*. Department of Water Resources, Victoria, Australia.

30. Gordon, N.D., McMahon, T.A., and Finlayson, B.L. (1992). *Stream Hydrology: An Introduction for Ecologists*. Wiley & Sons, New York, NY.

31. Simon, A. and Hupp, C.R. (1992). "Geomorphic and Vegetative Recovery Processes along Modified Stream Channels of West Tennessee." *U.S. Geological Survey Open File Report 91–502*, Nashville, TN.

32. Merigliano, M.F. (1997). "Hydraulic geometry and stream channel behavior: An uncertain link." *Journal of the American Water Resources Association*, 33(6), 1327–1336.

33. Fripp, J., Burns, M. and Caverly, J. (1998). "Stream stability assessment techniques." *Proceedings of the 1998 ASCE Wetlands Engineering River Restoration Conference*, Denver, CO.

34. Myers, T.J. and Swanson, S. (1992). "Variation of stream stability with stream type and livestock bank damage in Northern Nevada." *Water Resources Bulletin*, 28(4), 743–754.

35. Myers, T.J and Swanson, S. (1996). "Temporal and geomorphic variations of stream stability and morphology: Mahogany Creek, NV." *Water Resources Bulletin*, 32(2), 253–265.

36. Rosgen, D.L. (1994). "A classification of natural rivers." *Catena*, 22, 169–199.

37. Rosgen, D.L. (2001). *A Stream Channel Stability Assessment Methodology*. http://www.wildlandhydrology.com/assets/CHANNEL_STABILITY_.pdf (accessed May 4, 2006).

38. Copeland, R.R., McComas, D.N., Thorne, C.R., Soar, P.J., Jonas, M.M., and Fripp, J.B. (2001). "Hydraulic Design of Stream Restoration Projects." *Coastal and Hydraulics Laboratory Report* No. ERDC/CHL TR-01-28, U.S. Army Corps of Engineers, Washington, DC.

39. Annandale, G.W. (1994). *Guidelines for the Hydraulic Design and Maintenance of River Crossings. Volume VI: Risk Analysis of River Crossing Failure.* Department of Transport, Report No. TRH 25:1994, Pretoria, Republic of South Africa.

40. Annandale, G.W. (1999). "Risk analysis of river bridge failure." In *Stream Stability and Scour at Highway Bridges.* E.V. Richardson and P.F. Lagasse, Eds. ASCE, Reston, VA, 1003–1012.

41. Schumm, S.A. (1977). *Applied Fluvial Geomorphology*. Elsevier Publishing, Amsterdam, Netherlands.

42. Vermont Water Quality Division (2001). *Provisional Hydraulic Geometry Curves*. Vermont Department of Environmental Conservation, Waterbury, VT, www.vtwaterquality.org/rivers/docs/rv_hydraulicgeocurves.pdf.

43. Shields, Jr., F.D. (1996). *Hydraulic and Hydrologic Stability in River Channel Restoration: Guiding Principles for Sustainable Projects.* A. Brookes and F.D. Shields, Jr., Eds. Wiley & Sons, New York, NY.

44. Harvey, M.D. and Watson, C.C. (1986). "Fluvial processes and morphological thresholds in incised channel restoration." *Water Resources Bulletin,* 22(3), 359–368.

45. Thorne, C.R. and Osman, A.M. (1988). "Riverbank stability analysis II: Applications." *Journal of Hydraulic Engineering*, 114(2), 151–172.

46. Lawler, D.M., Thorne, C.R., and Hooke, J.M. (1997). "Bank erosion and instability." In *Applied Fluvial Geomorphology for River Engineering and Management.* C.R. Thorne, R.D. Hey, and M.D. Newson, Eds. Wiley & Sons, New York, NY.

47. Lewin, J., Macklin, M.G., and Newson, M.D. (1988). "Regime theory and environmental change—irreconcilable concepts?" In White, W.R. (Ed.), *International Conference on River Regime.* Wiley & Sons, Chichester, England, 431–445.

48. Chang, H.H. (1988). *Fluvial Processes in River Engineering*. Krieger Publishing Company, Melbourne, FL, 432 pp.

49. Bagnold, R.A. (1966). "An approach to the sediment transport problem from general physics." U.S. Geological Survey Professional Paper 422–I.

50. Brookes, A. (1989). *Channelized Rivers: Perspectives for Environmental Management.* Wiley & Sons, New York, NY.

51. Wilcock, P.R. (1997). "Entrainment, displacement and transport of tracer gravels." *Earth Surface Processes and Landforms*, 22(12), 1125–1138.

52. Wilcock, P.R. and Crowe, J.C. (2003). "Surface-based transport model for mixed-size sediment." *Journal of Hydraulic Engineering*, 129(2), 120–128.

53.   Parker, G., Klingeman, P.C., and McClean, D.G. (1982). "Bedload and size distribution in paved gravel-bed streams." *Journal of Hydraulic Engineering*, ASCE, 108(HY4), 544–571.

54.   Andrews, E.D. (1983). "Entrainment of gravel from naturally sorted riverbed material." *Geological Society of America Bulletin*, 94, 1225–1231.

55.   Carson, M.A. and Griffiths, G.A. (1987). "Bedload transport in gravel channels." *Journal of Hydrology*, 26, 1–151.

56.   Osman, A.M. and Thorne, C.R. (1988). "River bank stability analysis I: Theory." *Journal of Hydraulic Engineering*, 114(2), 134–150.

57.   Simon, A., Curini, A., Darby, S., and Langendoen, E.J. (1999). "Streambank mechanics and the role of bank and near-bank processes in incised channels." In *Incised River Channels* (Darby, S., and Simon, A., Eds.), Wiley & Sons, New York, NY, Chapter 6.

58.   ASCE Task Committee on Hydraulics, Bank Mechanics, and Modeling of River Width Adjustment (1998). "River width adjustment I: Processes and mechanisms." *Journal of Hydraulic Engineering*, 124(9), 881–902.

59.   Niezgoda, S.L. and Johnson, P.A. (2005). "Improving the urban stream restoration effort: identifying critical form and processes relationships." *Environmental Management*, 35(5), 579-592.

60.   Brice, J.C. and Blodgett, J.C. (1978). *Countermeasures for Hydraulic Problems at Bridges. Volume I. Analysis and Assessment.* U.S. Department of Transportation, FHWA, Report No. FHWA-RD-78-162, Washington, DC.

61.   Rosgen, D.L. (1996). *Applied River Morphology*. Wildland Hydrology, Pagosa Springs, CO.

62.   Fenneman, N.M. and Johnson, D.W. (1946). *Physical Division of the United States*. U.S. Geological Survey, scale 1:7,000,000. Available at http://tapestry.usgs.gov/physiogr/physio.html.

63.   Thornbury, W. D. (1965). *Regional Geomorphology of the United States*. Wiley & Sons, New York, NY.

64.   Dietz, R.S. (1952). "Geomorphic evolution of the continental terrace (continental shelf and slope)." *American Association of Petroleum Geologists Bulletin*, 36, 1802–1819.

65.   Atwood, W.W. (1940). *The Physiographic Provinces of North America*. Ginn and Company, New York, NY.

66.   Geological Survey (2003). *A Tapestry of Time and Terrain: The Union of Two Maps Geology and Topography*. http://tapestry.usgs.gov/physiogr/physio.html.

67.   Graf, W.L. (1987). *Geomorphic Systems of North America*. Geological Society of America, Boulder, CO.

68.   Mills, H.H., Brakenridge, G.R., Jacobson, R.B., Newell, W.L., Pavich, M.J., and Pomeroy, J.S. (1987). "Appalachian mountains and plateaus." In *Geomorphic Systems of North America*, Chapter 2. Geological Society of America, Boulder, CO.

69.   Hack, J.T. (1965). *Geomorphology of the Shenandoah Valley, Virginia and West Virginia, and Origin of the Residual Ore Deposits*. U.S. Geological Survey Professional Paper 484.

70.   Hack, J.T. (1957). *Studies of Longitudinal Stream Profiles in Virginia and Maryland*. U.S. Geological Survey Professional Paper 294–B, 45–97.

71. Maryland Department of Natural Resources (2002). *Maryland's Nonpoint Source Program Management Plan.* Department of Natural Resources, Annapolis, MD (available online at www.dnr.state.md.us/bay/czm/nps/plans/summary.html).

72. U.S. Geological Survey (2004). "Regionalization of channel geomorphology characteristics for streams in New York State." U.S. Geological Survey, Troy, NY.

73. Prestegaard, K.L. (2000). *Preliminary Assessment of Morphological and Hydrological Characteristics of Piedmont and Coastal Plain Streams in Maryland.* Report to Maryland Department of the Environment, Baltimore, MD.

74. Walker, H.J. and Coleman, J.M. (1987). *Coastal Plain, Geomorphic Systems of North America,* Chapter 3. Geological Society of America, Boulder, CO.

75. Sweet, W.V. and Geratz, J.W. (2003). "Bankfull hydraulic geometry relationships and recurrence intervals for North Carolina's Coastal Plain." *Journal of the American Water Resources Association*, 39(4), 861–871.

76. Osterkamp, W.R., Fenton, M.M., Gustavson, T.C., Hadley, R.F., Holliday, V.T., Morrison, R.B., Toy, T.J. (1987). "Great Plains." In *Geomorphic Systems of North America*, Chapter 6. Geological Society of America, Boulder, CO.

77. Doll, B.A., Wise-Frederick, D.E., Buckner, C.M., Wilkerson, D., Harman, W.A., Smith, R.E., and Spooner, J. (2002). "Hydraulic geometry relationships for urban streams throughout the Piedmont of North Carolina." *Journal of the American Water Resources Association*, 38(3), 641–652.

78. McCandless, T.L. and Everett, R.A. (2002). "Bankfull Discharge and Channel Characteristics of Streams in the Piedmont Hydrologic Region." *U.S. Fish and Wildlife Service*, Report CBFO-SO2-01.

79. Castro, J.M. and Jackson, P.L., 2001. "Bankfull discharge recurrence intervals and regional hydraulic geometry relationships." *Journal of the American Water Resources Association*, 37(5), 1249–1262.

80. Leopold, L.B. and Maddock, T. (1953). *The Hydraulic Geometry of Stream Channels and Some Physiographic Implications.* U.S. Geological Survey Professional Paper 252.

81. Osterkamp, W.R. and Hedman, E.R. (1982). "Perennial-streamflow characteristics related to channel geometry and sediment in Missouri River Basin." *U.S. Geological Survey Professional Paper 1242.*

82. Simon, A. and Rinaldi, M. (2000). "Channel instability in the loess area of the midwestern United States." *Journal of the American Water Resources Association*, 36(1), 133–150.

83. Madole, R.F., Bradley, W.C., Loewenherz, D.S., Ritter, D.F., Rutter, N.W., and Thorn, C.E. (1987). "Rocky Mountains." In *Geomorphic Systems of North America*, Chapter 7. Geological Society of America, Boulder, CO.

84. Dohrenwend, J.C. (1987). "Basin and Range." In *Geomorphic Systems of North America*, Chapter 9. Geological Society of America, Boulder, CO.

85. Muhs, D.R., Thorson, R.M., Clague, J.J., Mathews, W.H., McDowell, P.F., Kelsey, H.M. (1987). "Pacific Coast and Mountain System." In *Geomorphic Systems of North America*, Chapter 13. Geological Society of America, Boulder, CO.

86. Wilcock, P.R. and Kenworthy, S.T. (2002). "A two-fraction model for the transport of sand/gravel mixtures." *Water Resources Research*, 38(10), 121–133.